Springer Tracts in Advanced Robotics
Volume 26

Editors: Bruno Siciliano · Oussama Khatib · Frans Groen

Springer Tracts in Advanced Robotics

Edited by B. Siciliano, O. Khatib, and F. Groen

Vol. 25: Corke, P.; Sukkarieh, S. (Eds.)
Field and Service Robotics
580 p. 2006 [3-540-33452-1]

Vol. 24: Yuta, S.; Asama, H.; Thrun, S.; Prassler, E.; Tsubouchi, T. (Eds.)
Field and Service Robotics
550 p. 2006 [3-540-32801-7]

Vol. 23: Andrade-Cetto, J,; Sanfeliu, A.
Environment Learning for Indoor Mobile Robots
130 p. 2006 [3-540-32795-9]

Vol. 22: Christensen, H.I. (Ed.)
European Robotics Symposium 2006
209 p. 2006 [3-540-32688-X]

Vol. 21: Ang Jr., H.; Khatib, O. (Eds.)
Experimental Robotics IX
618 p. 2006 [3-540-28816-3]

Vol. 20: Xu, Y.; Ou, Y.
Control of Single Wheel Robots
188 p. 2005 [3-540-28184-3]

Vol. 19: Lefebvre, T.; Bruyninckx, H.; De Schutter, J.
Nonlinear Kalman Filtering for Force-Controlled Robot Tasks
280 p. 2005 [3-540-28023-5]

Vol. 18: Barbagli, F.; Prattichizzo, D.; Salisbury, K. (Eds.)
Multi-point Interaction with Real and Virtual Objects
281 p. 2005 [3-540-26036-6]

Vol. 17: Erdmann, M.; Hsu, D.; Overmars, M.; van der Stappen, F.A (Eds.)
Algorithmic Foundations of Robotics VI
472 p. 2005 [3-540-25728-4]

Vol. 16: Cuesta, F.; Ollero, A.
Intelligent Mobile Robot Navigation
224 p. 2005 [3-540-23956-1]

Vol. 15: Dario, P.; Chatila R. (Eds.)
Robotics Research – The Eleventh International Symposium
595 p. 2005 [3-540-23214-1]

Vol. 14: Prassler, E.; Lawitzky, G.; Stopp, A.; Grunwald, G.; Hägele, M.; Dillmann, R.; Iossifidis. I. (Eds.)
Advances in Human-Robot Interaction
414 p. 2005 [3-540-23211-7]

Vol. 13: Chung, W.
Nonholonomic Manipulators
115 p. 2004 [3-540-22108-5]

Vol. 12: Iagnemma K.; Dubowsky, S.
Mobile Robots in Rough Terrain – Estimation, Motion Planning, and Control with Application to Planetary Rovers
123 p. 2004 [3-540-21968-4]

Vol. 11: Kim, J.-H.; Kim, D.-H.; Kim, Y.-J.; Seow, K.-T.
Soccer Robotics
353 p. 2004 [3-540-21859-9]

Vol. 10: Siciliano, B.; De Luca, A.; Melchiorri, C.; Casalino, G. (Eds.)
Advances in Control of Articulated and Mobile Robots
259 p. 2004 [3-540-20783-X]

Vol. 9: Yamane, K.
Simulating and Generating Motions of Human Figures
176 p. 2004 [3-540-20317-6]

Vol. 8: Baeten, J.; De Schutter, J.
Integrated Visual Servoing and Force Control
198 p. 2004 [3-540-40475-9]

Vol. 7: Boissonnat, J.-D.; Burdick, J.; Goldberg, K.; Hutchinson, S. (Eds.)
Algorithmic Foundations of Robotics V
577 p. 2004 [3-540-40476-7]

Vol. 6: Jarvis, R.A.; Zelinsky, A. (Eds.)
Robotics Research – The Tenth International Symposium
580 p. 2003 [3-540-00550-1]

Vol. 5: Siciliano, B.; Dario, P. (Eds.)
Experimental Robotics VIII
685 p. 2003 [3-540-00305-3]

Vol. 4: Bicchi, A.; Christensen, H.I.; Prattichizzo, D. (Eds.)
Control Problems in Robotics
296 p. 2003 [3-540-00251-0]

Vol. 3: Natale, C.
Interaction Control of Robot Manipulators – Six-degrees-of-freedom Tasks
120 p. 2003 [3-540-00159-X]

Vol. 2: Antonelli, G.
Underwater Robots – Motion and Force Control of Vehicle-Manipulator Systems
209 p. 2003 [3-540-00054-2]

Vol. 1: Caccavale, F.; Villani, L. (Eds.)
Fault Diagnosis and Fault Tolerance for Mechatronic Systems – Recent Advances
191 p. 2002 [3-540-44159-X]

Geoffrey Taylor · Lindsay Kleeman

Visual Perception and Robotic Manipulation

3D Object Recognition, Tracking and Hand-Eye Coordination

With 96 Figures

Professor Bruno Siciliano, Dipartimento di Informatica e Sistemistica, Università degli Studi di Napoli Federico II, Via Claudio 21, 80125 Napoli, Italy, email: siciliano@unina.it

Professor Oussama Khatib, Robotics Laboratory, Department of Computer Science, Stanford University, Stanford, CA 94305-9010, USA, email: khatib@cs.stanford.edu

Professor Frans Groen, Department of Computer Science, Universiteit van Amsterdam, Kruislaan 403, 1098 SJ Amsterdam, The Netherlands, email: groen@science.uva.nl

Authors

Geoffrey Taylor
Lindsay Kleeman

Monash University
Intelligent Robotics Research Centre
Department of Electrical & Computer Systems Engineering
Monash University, VIC 3800
Australia

grtay1@yahoo.com.au
lindsay.kleeman@eng.monash.edu.au

ISSN print edition: 1610-7438
ISSN electronic edition: 1610-742X

ISBN-10 3-540-33454-8 **Springer Berlin Heidelberg New York**
ISBN-13 978-3-540-33454-5 **Springer Berlin Heidelberg New York**

Library of Congress Control Number: 2006923557

Springer is a part of Springer Science+Business Media
springer.com

Printed in Germany

Typesetting: Digital data supplied by authors.
Data-conversion and production: PTP-Berlin Protago-TEX-Production GmbH, Germany (www.ptp-berlin.com)
Cover-Design: design & production GmbH, Heidelberg
Printed on acid-free paper 89/3141/Yu - 5 4 3 2 1 0

STAR (Springer Tracts in Advanced Robotics) has been promoted under the auspices of EURON (European Robotics Research Network)

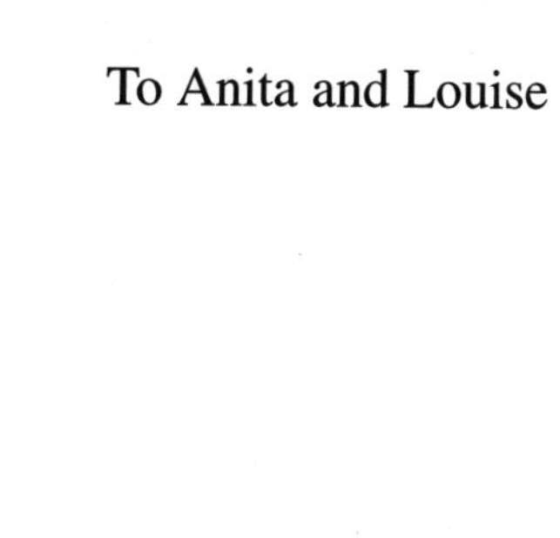

To Anita and Louise

Foreword

At the dawn of the new millennium, robotics is undergoing a major transformation in scope and dimension. From a largely dominant industrial focus, robotics is rapidly expanding into the challenges of unstructured environments. Interacting with, assisting, serving, and exploring with humans, the emerging robots will increasingly touch people and their lives.

The goal of the *Springer Tracts in Advanced Robotics (STAR)* series is to bring, in a timely fashion, the latest advances and developments in robotics on the basis of their significance and quality. It is our hope that the wider dissemination of research developments will stimulate more exchanges and collaborations among the research community and contribute to further advancement of this rapidly growing field.

The monograph written by Geoffrey Taylor and Lindsay Kleeman is an evolution of the first Author's Ph.D. dissertation. This book provides an integrated systems view of computer vision and robotics, covering a range of fundamental topics including robust and optimal sensor design, visual servoing, 3D object modelling and recognition, and multi-cue tracking. The treatment is in-depth, with details of theory, real-time implementation, and extensive experimental results. As such, the book has wide appeal to both theoretical and practical roboticists. Furthermore, the culmination of the work with the demonstration of two real-world domestic tasks represents a first step towards the realization of autonomous robots for service applications.

The first book in the series to be accompanied by a comprehensive multimedia supplement, this title constitutes a very fine addition to STAR!

Naples, Italy,
February 2006

Bruno Siciliano
STAR Editor

Preface

Autonomous robots have the potential to revolutionize the service industry. Specialized applications such as palletizing, vacuum cleaning and tour guiding can be solved with today's technology, but next-generation universal robots will require new solutions to the challenge of operating in an unconstrained human environment. For example, imagine you suffer from arthritis so severe that your daily needs go unattended. A robotic aid could restore your independence by helping to put on your shoes, pour a drink, stack dishes, put away groceries or a host of other household tasks involving lifting, bending and carrying. Such a robot must be capable of interpreting and planning actions based on simple supervisory commands, recognizing a variety of objects despite a continuous spectrum of variation in appearance, and manipulating a dynamic environment in continuously new ways. Unlike their industrial counterparts, widespread adoption of domestic robots demands minimal reliance on application specific knowledge and high robustness to environmental changes and inevitable operational wear. Rich sensing modalities such as vision are therefore likely to play a central role in their success. This book takes steps towards the realization of domestic robots by presenting the fundamental components of a 3D model-based robot vision framework. At all stages from perception through to control, emphasis is placed on robustness to unknown environmental conditions and calibration errors.

At the lowest level of perception, stereoscopic light stripe scanning captures a dense range map of the environment. Unlike conventional structured light techniques, stereoscopic scanning exploits redundancy to robustly identify the primary light stripe reflection despite secondary reflections, cross-talk and other sources of interference. An automatic procedure to calibrate the system from the scan of an arbitrary non-planar object is also described.

At the next level of perceptual abstraction, range data is segmented and fitted with geometric primitives including planes, cylinders, cones and spheres. Central to this process is a surface type classification algorithm that characterizes the local topology of range data. The classifier presented in this book is shown to achieve significantly greater noise robustness than conventional techniques. Many classes of domestic objects can be identified as composites of extracted primitives using a graph matching technique, despite significant variations in appearance.

At the highest level of perception, model-based tracking compensates for scene dynamics and calibration errors during manipulation. Selection of salient tracking features is a challenging problem when objects and visual conditions are unknown. This book explores the use of multi-cue fusion based on edge, texture and colour cues, which achieves long-term robustness despite losing individual cues as objects and visual conditions vary. Multi-cue tracking is shown to succeed in challenging situations where single-cue trackers fail.

Finally, robust hand-eye coordination to perform useful actions requires visual feedback control of a robot manipulator. This book introduces hybrid position-based visual servoing, which fuses kinematic and visual measurements to robustly handle occlusions and provide a mechanism for on-line compensation of calibration errors, both classical problems in position-based visual servoing.

The results of extensive testing on an upper-torso humanoid robot are presented to support the framework described in this book. The culmination of the experimental work is the demonstration of two real-world domestic tasks: locating and grasping an unknown object and pouring the contents of an interactively selected cup into a bowl. Finally, the role of vision in a larger multi-sensor framework is explored through the task of identifying a cup of ethanol from among several candidates based on visual, odour and airflow sensing.

Accompanying this book is a CD-ROM containing video clips, VRML data, C++ code with data sets and lecture slides to supplement the discussions and experimental results. This material is referred to as the *Multimedia Extensions* at relevant points in the text. In addition, video clips of preliminary and related experiments not discussed in main text are also presented, providing a background to the development of this work. To view the *Multimedia Extensions*, insert the CD-ROM in a PC disk drive and a web browser should launch with the HTML interface. If the browser does not launch automatically, the interface can be accessed by opening `index.html` in the root directory of the CD-ROM.

The authors would like to thank R. Andy Russell, who opened up interesting new research directions by collaborating on the odour and airflow sensing experiments in Chapter 7, and Sven Molin and Åke Wernersson for making their stay at Luleå University of Technology, Sweden, a valuable experience and for motivating the light stripe research in Chapter 3. This research was supported by an *Australian Postgraduate Award* and the *Strategic Monash University Research Fund* (SMURF) for the *Humanoid Robotics: Perception, Intelligence and Control* project at the *Intelligent Robotics Research Centre* (IRRC), and the *Australian Research Council Centre for Perceptive and Intelligent Machines in Complex Environments* (PIMCE). The funding is gratefully acknowledged.

Monash University,
February 2006

Geoffrey Taylor
Lindsay Kleeman

This book is the result of four years of postgraduate research at the Department of Electrical and Computer Systems Engineering, Monash University. Thank you to those staff and students who helped both directly and indirectly to make my journey a colourful, productive and enjoyable one. I am particularly grateful to my postgraduate supervisor and co-author Lindsay Kleeman, whose extreme generosity with time, encouragement and expertise were both a benefit to this book and my personal development as a research scientist. This book would not have been possible without the patience and support of my parents, Ron and Jeni Taylor, and my wife Anita, who provided the stable environment that allowed me to pursue my dreams.

February 2006 *Geoffrey Taylor*

Contents

List of Figures

List of Tables

List of Acronyms

CAD	Computer Aided Design
CCD	Charge-Coupled Device
DOF	Degrees of Freedom
ECL	Endpoint Closed-Loop
EGI	Extended Gaussian Image
EOL	Endpoint Open-Loop
IEKF	Iterated Extended Kalman Filter
LED	Light Emitting Diode
LM	Levenberg-Marquardt
MMX	Multimedia Extensions
PCA	Principal Components Analysis
RGB	Red Green Blue
ROI	Region Of Interest
SIMD	Single Instruction, Multiple Data
SSD	Sum of Squared Difference
SSE	Streaming SIMD Extensions
VRML	Virtual Reality Markup Language

1

Introduction

If the early parallels between computers and robots hold up, we can expect to see ... killer applications for robots in the next ten to fifteen years, and by the year 2020 robots will be pervasive in our lives.
– Rodney Brooks, Flesh and Machines

Since the middle of last century, technologists have confidently predicted the imminent revolution in domestic and service robotics. We are told that intelligent robots will begin to appear in guises such as cars, vacuum cleaners, and even humanoids. These machines will lift the burden of work in our homes and offices by taking the role of cleaner, courier, nurse, and security guard among others. Robots will recognize their owners by appearance and voice, and understand natural modes of communication including gestures and speech. We will give simple commands such as *"Set the table"* or *"Bring the coffee to my desk"*, and our service robot will have the necessary sensing and intelligence to fulfill the request. Robotic tele-presence will allow owners to remotely feed pets, check security and perform other tasks while away. Robotic aids will improve the quality of life for the elderly and disabled, while society as a whole will benefit from the reduced demand on health care services. Eventually, service robots will roam freely around our homes and offices to help in whatever way they see fit, driven by an internal emotional model and learning from their surroundings.

While the above scenario is still a fantasy, the steady decrease in the size and cost of sensors and computing resources over the past decade has made the possibilities of service robots seem even closer and more compelling than ever. Robots are sure to find widespread appeal as they continue the trend in home automation that has been underway for over a century. Appliances such as sewing machines, washing machines, central heating and kettles are now ubiquitous in our lives. All contain what may be described in robotic terms as sensors, actuators and signal processors, and can adapt their operation in a limited sense to environmental conditions. More importantly, each new technology is readily accepted as a positive contribution to the standard of living. The primitive precursors of domestic robots have already found a place the home.

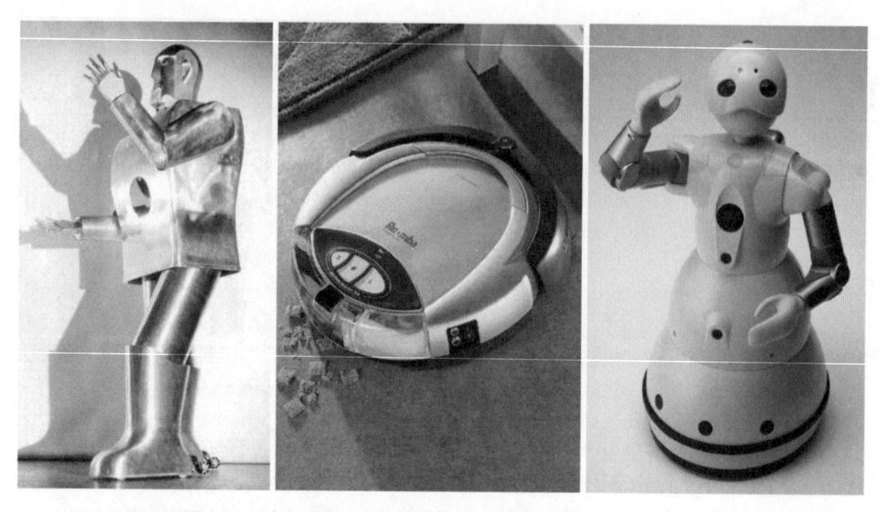

Fig. 1.1. Service robots of the past, present and future. *Elektro* (left) was built by Westinghouse in 1937 and exhibited as the ultimate appliance, although it was little more than a complex automaton. Present-day service robots such as the *Roomba* (centre) from iRobot can perform specific tasks such as vacuum cleaning autonomously, but have limited application. The concept robot *Wakamaru* (right) from Mitsubishi is designed to develop its own motivations to help in the daily life of its user, and represents the first generation of universal aids. *Elektro image from http://davidszondy.com/future/robot/elektro1.htm. Roomba image from http://www.frc.ri.cmu.edu/ hpm/talks/Extras/roomba.JPG. Wakamaru image from http://www.expo2005.or.jp/jp/C0/C3/C3.7/C3.7.5/.*

Figure 1.1 provides a perspective on the evolution and continuing trends in domestic service robotics. Present-day commercial service robots generally take the form of autonomous vacuum cleaners and lawn mowers. These robots are equipped with a variety of sensors that are used to avoid obstacles, seek out re-charging stations, and even perform simultaneous localization and mapping in an unknown environment. While these capabilities represent significant strides in home automation, modern cleaning robots, like refrigerators and washing machines, are still highly specialized devices that do not deliver the promise of a universal aid.

In both research and popular culture, humanoid robots have long been regarded as the ultimate universal aid. The reason is simple: our offices and homes are naturally designed for humans, and robots in humanoid form can effectively exploit existing infrastructure. The humanoid form also facilitates human-machine interaction, allowing people to communicate and work cooperatively and intuitively with their robotic companions. Our knowledge and intuition about how humans reason and interact with the world will boot-strap the development of parallel humanoid robot capabilities. Any activity that can be performed by a person, such as climbing a ladder and squeezing through a man hole, can (in principle at least!) be mimicked by a humanoid robot. Unhindered by the constraints of biology, humanoid robots will also develop "super-human" skills. Sensory information such as sonar, structured light, night vision, or cameras on each fingertip will usher as yet unimagined possibilities and challenge our view of human-like intelligence.

In anticipation of this ultimate trend in home automation, humanoid robots have recently started to progress from academia into the commercial world. HOAP-1 is a small-scale (48 cm tall) humanoid robot developed by Fujitsu[1] and marketed towards researchers and developers. In contrast, Sony's more recent Qrio[2] is described as an entertainment robot. The ambitious medium-scale wheeled humanoid robot known as Wakamaru from Mitsubishi[3] is intended to provide practical domestic services such as security, health care, access to information and tele-presence. Toyota's Partner Robots[4] are intended to provide similar functionality. Along with the more well known humanoid robots such as Asimo from Honda Motor and HRP-2 from Kawada Industries, these robots may well represent the first practical universal aids.

Just as humanoid robots are inspired by familiar forms, the importance of vision in human perception will likely be mirrored in our service robots. Physical coordination, navigating through complex environments and recognizing objects and people are all enhanced by visual information. We take these skills for granted, yet much research remains to be done before we grasp the underlying mechanisms. The science of computer vision aspires to emulate our ability to extract meaningful information from images of the world. Taking this goal a step further, robot vision explores how visual information can be used to drive interactions with the world, by linking perception to action. Robot vision has already been an active area of research for several decades. Limitations on computing power resulted in early robotic vision systems being either slow, like Shakey the Robot from Stanford Research Institute[5], large, such as the early Navlab vehicles from Carnegie Mellon University[6], or tethered to off-board computing, such as the keyboard-playing Wabot-2 from Waseda University[7]. However in more recent years, the exponential increase in available computing power and decrease in the cost of both computers and cameras has stimulated an acceleration in our understanding of robot vision. Compact, untethered robots that use real-time visual sensing to navigate complex environments, locate and grasp objects and interact with humans are now a reality.

Three things are almost certain about universal service robots of the future: many will have manipulators (and probably legs!), most will have cameras, and almost all will be called upon to grasp, lift, carry, stack or otherwise manipulate objects in our environment. Visual perception and coordination in support of robotic grasping is thus a vital area of research for the progress of universal service robots. With their imminent arrival, the time is ripe to direct research effort towards the skills that will facilitate widespread application.

---

[1]http://www.automation.fujitsu.com/en/products/products07.html

[2]http://www.sony.net/SonyInfo/QRIO/top_nf.html

[3]http://www.mhi.co.jp/kobe/wakamaru/english/

[4]http://www.toyota.co.jp/en/special/robot/

[5]http://www.sri.com/about/timeline/shakey.html

[6]http://www.cs.cmu.edu/afs/cs/project/alv/www/index.html

[7]http://www.humanoid.rise.waseda.ac.jp/booklet/kato02.html

1.1 Motivation and Challenges

The development of general-purpose service robots will require advances in diverse fields including mechanics, control, sensing and artificial intelligence. Significant progress has already been made in many areas including stable bipedal locomotion on uneven terrain [144, 145], the related problems of localization and navigation [22, 118], mechanical compliance for safe interaction [119], dextrous manipulation [95, 103], social interaction [14, 78], various sensing and actuation modalities and a range of problems in artificial intelligence. The focus of this book is on identifying and manipulating unknown objects under visual guidance, which is a fundamental skill for practical service applications such as assisting the elderly and disabled. Object recognition and manipulation are classical problems in robotics, and many techniques already exist for handling unknown objects. However, much work remains to be done in extending these techniques to address the complexities and uncertainties of an unprepared domestic environment. This book aims to address this need by providing a framework for visual perception and manipulation of unknown objects, with an emphasis on robustness for autonomous service robots in the real world.

The functional building blocks that enable a robot to perform arbitrary tasks can be broadly divided into the four areas of human-machine interaction, perception, intelligence and control. Bidirectional human-machine interaction is necessary to communicate the requirements of a task to the robot and relate the progress back to the user. Interaction with service robots will likely to take the form of natural speech and gestures, or a simple graphical interface in the case of remote operation. Perception is an essential building block for autonomous behaviour and encompasses both low-level sensing and high-level abstraction of useful information about the world. For robotic vacuum cleaners and similar devices, perception may simply drive reactive behaviours without intervening intelligence. Complex tasks may require a more sophisticated approach involving the maintenance of a consistent world model, and associated high-level interpretation to drive planning, prediction and monitoring of actions. These latter activities are the function of intelligence, which provides the link between interaction, perception and control.

Robots face many new and complicating challenges when promoted from industrial to domestic applications. Interaction must be intuitive and robust to ambiguous interpretation. Task specifications are likely to be *ad hoc* and therefore incomplete as not all pertinent information is available. The desired behaviour must be either inferred by the robot or actively pursued by prompting for further interaction. The robot may be required to learn new behaviours and condition existing behaviour based on feedback from the user. Sensing must provide useful measurements over a wide variety of unpredictable operating conditions. The robot must be capable of identifying and locating broad classes of objects in cluttered surroundings. Control of the limbs must be robust to wear and other incidental variations in mechanical parameters. All of these challenge must be met with real-time operation if the robot is to interact naturally with humans.

Addressing all of these challenges is beyond the scope of this book. We therefore focus on the core building blocks of visual perception and control, which nevertheless provide the foundation for domestic tasks involving more sophisticated intelligence and interaction. Within this limited scope, four main areas of difficulty can be identified: imprecise information about tasks and objects, operation in an unstructured, cluttered environment, robustness to uncertainty in calibration, and real-time operation. The working hypothesis is that these specific challenges can be satisfied in a framework based on visual sensing, in the same manner that vision is the primary sense for humans. These main challenges are elaborated below:

Imprecise task specifications and lack of prior knowledge

Conventional industrial robots are characterized by repetitive tasks in a precisely calibrated workspace. At the other extreme, almost every domestic task differs with respect to the location and identity of objects, and the required manipulations. Domestic tasks are usually only specified at a supervisory level with reference to general actions and classes of objects, as in the simple example: "Please get the cup from the table." Lack of prior knowledge affects all stages of sensing, planning and actuation. Firstly, the robot must be able to recognize and locate a cup without necessarily having seen the object before. Visual sensing cannot assume the presence of suitable colours and textures for accurate measurements. Segmentation, the process of dividing the visual data into meaningful regions, is confounded by the distinction between adjacent unknown objects and different parts of the same unknown object. The unknown cup must be distinguished from a soft drink can, honey pot or other similarly shaped objects that may be present in the scene. Furthermore, the existence of a cup is not guaranteed if the command was not given in the vicinity of the table. If the cup is found, the robot must plan and execute a stable grasp without having previously handled the object. If the cup is not found or the results are ambiguous, the robot may need to request additional information based on existing sensory data, such as: "Do you mean the red or blue cup?" Dealing with imprecise knowledge about the required task drives many design choices in the framework presented in this book.

Robust visual sensing in a cluttered environment

Data association is required in almost all sensing applications, and is the general problem of determining the identity of primitives extracted from sensor data by relating features to each other and/or an internal world model. Association errors can lead to false internal models and, ultimately, catastrophic failure. Any sensing application can suffer from association errors, and the opportunity for error is particularly acute in cluttered, dynamic, unpredictable environments such as encountered by domestic service robots. Visual sensing can be hindered by background colours and textures, lighting variation, reflections, image noise and a variety of other distractions. Robust data association is thus a significant challenge in service applications.

Another effect of operation in a cluttered, unpredictable environment is the possibility of occlusions and the resulting loss of visual information. This can impact the ability to recognize objects, update the world model and control limbs using hand-eye coordination. Perception and control algorithms with high tolerance to occlusions are therefore desirable.

Robustness to operational wear and calibration errors

In the classical solution to robotic reaching and grasping, visual sensing accurately recovers the metric position of the object, and the end-effector is accurately driven to the desired location using a kinematic model. Any errors in the calibration of the sensors or mechanics, which becomes more likely as the robot increases in complexity, can lead to failure of this simple approach. Two sources of variation are readily identified: operational wear, and the complexity of calibration. Industrial robots also suffer from these problems, but the effects are amplified in domestic applications. Accidents are a real possibility in an unpredictable environment and accelerate the process of wear. Constraints on the weight and safety of service robots require light, compliant structures that are difficult to model kinematically. To find widespread application, robots must be produced with cheaper sensors and relaxed manufacturing tolerances. They will also need to operate autonomously for extended periods without costly technical maintenance, while tolerating the effects of wear on sensors and mechanical components. Overall, the development of perception and control methods that reduce the reliance on accurate calibration may ultimately lead to robots that are cheaper, more reliable and require lower maintenance.

Real-time operation

To interact and work cooperatively with humans, service robots must be capable of operating with comparable reflexes, that is, in real-time. Industrial robots are characterized by super-human speed and accuracy. Conversely, visual sensing and planning algorithms often perform many times slower than a human in tasks such as scene interpretation and grasp planning. This problem is usually solved using off-board networks of parallel computers and specialized signal processing hardware. Unfortunately, the general computing resources available to a practical service robot are bounded by physical dimensions and power consumption and therefore likely to be much lower. Some specialized hardware may be present, such as "smart" cameras with simple image processing capabilities. In the long term, the constraints of real-time operation will progressively weaken as general purpose computers continue to increase in speed. Real-time operation guides the development of many algorithms presented in this book, but is usually secondary to the considerations described above.

1.2 Chapter Outline

The framework presented in this book addresses all stages of the *sense-plan-act* cycle for autonomously manipulating unknown objects in a domestic environment. Figure

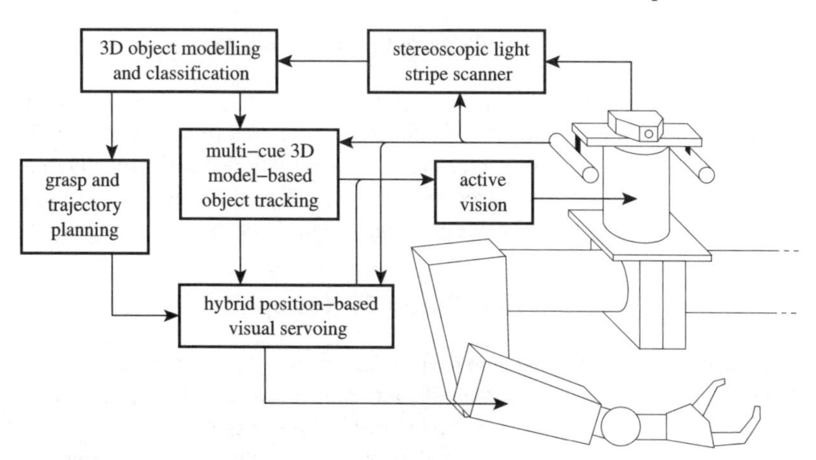

Fig. 1.2. Block diagram of proposed framework for classifying and manipulating unknown objects with robustness to uncertainties in a domestic environment.

1.2 illustrates the structure of this framework and serves as a road-map to the concepts and techniques developed in later chapters.

In the proposed framework, perception of the world begins with the *stereoscopic light stripe scanner* and *object classification and modelling* blocks working together to generate data-driven, textured, polygonal models of segmented objects. Object modelling and classification provides the link between low-level sensing and high-level planning. Unknown objects are classified into families such as cups, bowls or boxes based on geometry. This allows the robot to perform *ad hoc* tasks with previously unknown objects without a large database of learned models. Colour and texture information enables objects in the same family to be distinguished from each other. The perceived world model is maintained using *multi-cue 3D model-based tracking* to continuously estimate the state of each object based on texture, colour and geometric measurements. Tracking is necessary even for static objects to compensate for camera motion and detect collisions or unstable grasps. Finally, the desired manipulation is performed by controlling the end-effector using visual feedback in a *hybrid position-based visual servoing* framework. During servoing, *active vision* controls the gaze direction to maximize available visual information.

The following chapters fill in the details of the framework sketched above, starting from low-level sensing through to high-level perception and control, the finally concluding with experimental results to illustrate the integration of the entire framework. A brief outline of each chapter is presented in the following sections. In addition, the accompanying *Multimedia Extensions* CD-ROM provides videos, VRML data, C++ code and data sets to supplement this monograph.

Chapter 2: Foundations of Visual Perception and Control

Chapter 2 establishes a foundation of concepts, notational conventions and system models that will form the basis of algorithms and analyses in later chapters. Topics covered include homogeneous vectors and transformations, the projective camera model, kinematics of an active stereo camera head, digital image processing, and a survey of related work in visual perception and control for robotic manipulation.

Chapter 3: Shape Recovery Using Robust Stereoscopic Light Stripe Scanning

Chapter 3 describes the *stereoscopic light stripe scanner* block in figure 1.2. Acquiring depth information is a fundamental first step in building the 3D world model used to locate and classify objects of interest. Light stripe scanning is an active triangulation-based ranging technique that provides greater accuracy than related passive methods and has been used in robotics and computer vision for several decades. However, almost all conventional scanners are designed to operate under controlled conditions and cannot cope with the reflections, clutter, cross-talk or objects with unknown geometries, textures and surface properties that characterize a service robot's environment. This chapter reviews conventional light stripe scanning and then shows how a robust stereoscopic scanner can be designed to overcome these problems. Robustness is achieved by exploiting redundancy and geometric constraints to identify true measurements in the presence of noise. Validation, 3D reconstruction and auto-calibration techniques are derived, and implementation details including image processing are presented. Experimental results validate the system model and demonstrate the improvement over conventional techniques.

Chapter 4: 3D Object Modelling and Classification

To minimize prior knowledge, a service robot would benefit from the ability to classify unknown objects into known families such as cups, plates and boxes. The classification framework presented in Chapter 4 is based on modelling objects as collections of geometric primitives (planes, cylinders, cones and spheres). The primitives are extracted automatically from range data using a segmentation algorithm based on *surface type classification*. The classifier uses principal curvatures (from Gaussian image analysis) and convexity to determine the local shape of range patches. Geometric primitives are fitted to connected segments of homogeneous surface type, and an iterative refinement attempts to merge as many segments as possible without degrading the fitted models. Finally, the algorithm constructs an attributed graph expressing the relationship between segmented primitives, and families of objects are recognized using sub-graph matching. The segmentation, surface type classification, model-fitting and graph matching algorithms are described in detail, and experimental results are presented for several scenes. In particular, the non-parametric surface type classifier described in this chapter is shown to achiever greater robustness to noise than conventional classifiers without additional computational expense.

Chapter 5: Multi-cue 3D Model-Based Object Tracking

To close the visual feedback loop, a 3D model-based tracking algorithm continuously refines the pose of objects extracted from range data. Model-based tracking algorithms typically choose a particular cue such as edges or texture to locate the tracked target in a captured video. Unfortunately, the failure mode of any particular cue is almost certain to arise at some time in the unpredictable lighting and clutter of a domestic environment, making single-cue trackers unsuitable for service robots. Chapter 5 shows how multi-cue 3D model-based tracking overcomes this problem based on the notion that different cues exhibit complementary failure modes and are thus unlikely to fail simultaneously. The multi-cue tracker fuses intensity edges, texture and colour cues from stereo cameras in an extensible Kalman filter framework. Image processing, feature association and measurement models for each cue are described in detail, along with the Kalman filter implementation. Experimental results demonstrate the robustness of multi-cue 3D model-based tracking in the presence of visual conditions that otherwise cause single-cue trackers to fail, including low contrast, lighting variations, motion blur and occlusions.

Chapter 6: Hybrid Position-Based Visual Servoing

Chapter 6 addresses the problem of robust visual control of robotic manipulators for service applications. Traditional *look then move* control requires accurate kinematic and camera calibration, which is unrealistic for service robots. *Position-based visual servoing* is a feedback control technique in which the end-effector and target are continuously observed to minimize the effect of kinematic uncertainty. However, conventional position-based visual servoing is limited by a reliance on accurate camera calibration for unbiased pose estimation, and fails completely when the end-effector is obscured. This chapter presents an error analysis of the camera model in visual servoing to show that the pose bias can be corrected by introducing a simple scale factor. Furthermore, the problem of visual obstructions can be overcome by fusing visual and kinematic measurements in a new framework called *hybrid position-based visual servoing*. Kinematic measurements allow servoing to continue when the end-effector is obscured, while visual measurements minimize the uncertainty in kinematic calibration. A Kalman filter framework for estimating the visual scale factor and fusing visual and kinematic measurements is described, along with image processing, feature association and other implementation details. Experimental results demonstrate the improved accuracy and robustness of the hybrid approach compared to conventional servoing.

Chapter 7: System Integration and Experimental Results

In Chapter 7, the components developed in previous chapters are integrated into a complete *sense-plan-action* framework for service robots. The complete framework is implemented on an experimental upper-torso humanoid robot, and three simple domestic tasks are designed to evaluate the performance of the system. The first task requires the robot to fetch an object specified only by colour and family, the second task requires the robot to grasp an interactively selected cup and pour the contents into a bowl, and the final task requires the robot to identify and grasp a cup of ethanol based on airflow and chemical sensor readings. Implementation details of the humanoid robot, including simple human-machine interaction, grasp planning and task planning components are provided.

Appendices

Appendix A describes the mechanical properties and calibration procedure for the Biclops active stereo head on the humanoid service robot used for experiments in this book.

Appendix B details the analytical models used in the development of the stereoscopic light stripe scanner described in Chapter 3.

Appendix C provides details of the Iterated Extended Kalman Filter used by the model-based tracking algorithms developed in Chapters 6 and 5.

Appendix D details the stereo reconstruction error models used in the development of the hybrid position-based visual servo controller in Chapter 6.

Appendix E describes a simple calibration procedure to determine the various acquisition and actuation delays between inputs and outputs of the system.

Appendix F details the simple task planning algorithms used in the implementation of the experimental service robot in Chapter 7.

2

Foundations of Visual Perception and Control

The purpose of this chapter is to provide a brief overview of robotic and computer vision concepts, algorithms and mathematical tools that form the basis of the framework developed in the remainder of this book. Section 2.1 lays the groundwork with a brief discussion of homogeneous vectors, transformations, and representations of rotation and velocity typically used to model robotic and computer vision systems. Section 2.2 builds on this foundation with descriptions of models for the common components of a 3D robotic vision system, namely the imaging sensor and active stereo camera head. Finally, Sections 2.3 and 2.4 paint a broad outline of issues and methods in visual perception and control as they relate to robotic manipulation. In particular, we discuss how the choice of representation in perceptual models affects the capabilities of a visual system, and provide a survey of recent work in visual servoing.

2.1 Mathematical Preliminaries

2.1.1 Homogeneous and Inhomogeneous Vectors

Most geometric models in robotic vision systems involve transformations within and between three dimensional Euclidean space $\mathbb{R}^3$ and the two dimensional space $\mathbb{R}^2$ of the image plane. To distinguish these spaces, we adopt the convention of labelling vectors in $\mathbb{R}^3$ using upper-case bold symbols such as $\mathbf{X}$ and vectors in $\mathbb{R}^2$ using lower-case bold, as in $\mathbf{x}$. Analytical models can be cumbersome in Euclidean space since rotations, translations and camera projections are described by different linear and non-linear transformations. Under the alternative framework of *projective geometry*, all coordinate transformations are unified as linear matrix multiplications, known as *homogeneous* or *projective transformations* (see next section). Projective geometry is therefore used extensively in place of Euclidean geometry to model robotic and computer vision systems.

To apply homogeneous transformations, a point in $\mathbb{R}^n$ is replaced by an equivalent *homogeneous* $(n+1)$-vector in the projective space $\mathbb{P}^n$. The projective space

$\mathbb{P}^n$ is defined as the space of linear subspaces of $\mathbb{R}^{n+1}$. Consider a point in $\mathbb{R}^3$ with inhomogeneous coordinates X, Y, and Z; by introducing a scalar λ, a corresponding linear subspace of points $\mathbf{X}(\lambda)$ can be defined in $\mathbb{R}^4$ as

$$\mathbf{X}(\lambda) = \lambda(X,Y,Z,1)^\top \tag{2.1}$$

which is in fact a ray emanating from the origin in the direction of $(X,Y,Z,1)^\top$. Equation (2.1) specifies the homogeneous coordinates of the point at X, Y, and Z in the projective space $\mathbb{P}^3$. By convention and without loss of generality, the unknown scale can be discarded and the homogeneous coordinates written as $\mathbf{X} = (X,Y,Z,1)^\top$, which will be the convention adopted unless otherwise specified. An alternative convention sometimes encountered in the literature is to write $\mathbf{X} \simeq (X,Y,Z,1)^\top$, where $\simeq$ denotes equivalence only up to the unknown scale λ. For the image plane vector $\mathbf{x}$ in $\mathbb{R}^2$, the corresponding homogeneous coordinates in $\mathbb{P}^2$ can be written as $\mathbf{x} = (x,y,1)^\top$. In Section 2.2.1, we will see how expressing image coordinates in this way reduces a non-linear Euclidean camera model to a linear projective transformation.

Just as each inhomogeneous vector can be associated with a homogeneous representation, we can also define the inverse mapping. For a general homogeneous vector $\mathbf{X} = (X,Y,Z,\lambda)^\top$ with $\lambda \neq 0$, the equivalent Euclidean point $\mathbf{X}'$ is defined as

$$\mathbf{X}' = (X',Y',Z')^\top = (X/\lambda, Y/\lambda, Z/\lambda)^\top \tag{2.2}$$

When $\lambda = 0$, the homogeneous vector $\mathbf{X} = (X,Y,Z,0)^\top$ can be interpreted loosely as a point "at infinity" in the direction of the inhomogeneous vector $(X,Y,Z)^\top$.

Since the homogeneous and inhomogeneous coordinates of a point are equivalent, we do not distinguish these representations with different notation. The representation used in a given analytical model can generally be inferred from context, for example when a homogeneous transformation is applied to a point. An overview of projective geometry for computer vision can be found in [36] and [52], and a detailed discussion of homogeneous coordinates and transformations for robotics can be found in [102].

2.1.2 Coordinate Frames and Homogeneous Transformations

Coordinate vectors, whether homogeneous or inhomogeneous, must be specified with respect to a frame that defines the point of origin and orientation of the canonical axes. Coordinate frames are denoted using uppercase italics such as F. The frame in which a particular point is measured is appended as a superscript, such as ${}^F\mathbf{X}$, but may be omitted when the frame is clear from the context. Coordinates are transformed from one frame to another by application of a *homogeneous transformation* (or *homography*). Let ${}^A\mathbf{X}$ and ${}^B\mathbf{X}$ represent the location of a single point with respect to frames A and B, as shown in figure 2.1. Let the inhomogeneous vector ${}^B\mathbf{T}_A$ specify the position of the origin of frame A in frame B. The orientation of frame A with respect to frame B can be defined as a 3×3 rotation matrix ${}^B\mathbf{R}_A$ (see also the discussion in the following section). Then, the transformation of the homogeneous coordinates

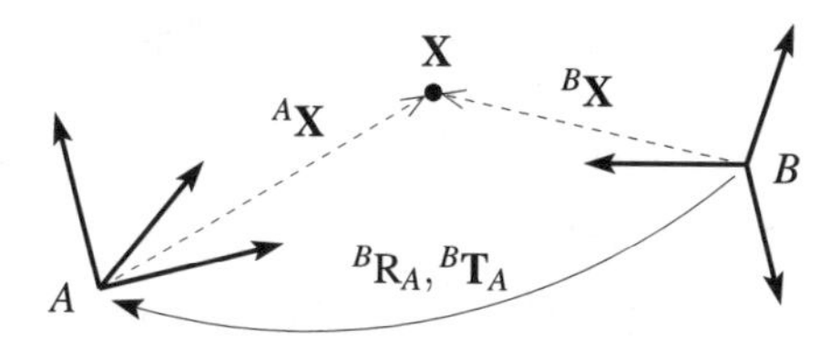

Fig. 2.1. Transformation of a point between coordinate frames.

of $\mathbf{X}$ from A to B can be written as ${}^B\mathbf{X} = {}^B\mathrm{H}_A{}^A\mathbf{X}$, where the 4×4 homogeneous transformation matrix ${}^B\mathrm{H}_A$ is given by:

$$ {}^B\mathrm{H}_A = \begin{pmatrix} {}^B\mathrm{R}_A & {}^B\mathbf{T}_A \\ \mathbf{0}_{1\times 3} & 1 \end{pmatrix} \tag{2.3} $$

Matrix ${}^B\mathrm{H}_A$ may be interpreted as the coordinate transformation from frame A to B or alternatively as the *pose* (orientation and position) of frame A with respect to frame B; both interpretations will be called upon in later formulations.

2.1.3 Representation of 3D Orientation

Equation (2.3) introduced the 3×3 rotation matrix ${}^B\mathrm{R}_A$ as one possible representation of 3D orientation. However, the rotation matrix representation over-parameterizes 3D orientation with 9 parameters for 3 degrees of freedom. In practice, this can make rotation matrices cumbersome to use in analytical models and also lead to numerical instabilities. Two common alternative representations of 3D orientation are *Euler angles* and *quaternions*. Euler angles specify a triplet of canonical rotations about the axes of a coordinate frame. Numerous Euler angle conventions regarding the order and axes of rotation are found in the literature; this book adopts the *yaw-pitch-roll* convention which specifies rotations by ϕ, θ and ψ about the X, Y and Z-axes in order. The equivalent rotation matrix can be calculated as

$$ \begin{aligned} \mathrm{R}(\phi,\theta,\psi) &= \mathrm{R}_z(\psi)\mathrm{R}_y(\theta)\mathrm{R}_x(\phi) \\ &= \begin{pmatrix} \cos\psi & -\sin\psi & 0 \\ \sin\psi & \cos\psi & 0 \\ 0 & 0 & 1 \end{pmatrix} \begin{pmatrix} \cos\theta & 0 & \sin\theta \\ 0 & 1 & 0 \\ -\sin\theta & 0 & \cos\theta \end{pmatrix} \begin{pmatrix} 1 & 0 & 0 \\ 0 & \cos\phi & -\sin\phi \\ 0 & \sin\phi & \cos\phi \end{pmatrix} \\ &= \begin{pmatrix} c_\theta c_\psi & s_\phi s_\theta c_\psi - c_\phi s_\psi & c_\phi s_\theta c_\psi + s_\phi s_\psi \\ c_\theta s_\psi & s_\phi s_\theta s_\psi + c_\phi c_\psi & c_\phi s_\theta s_\psi - s_\phi c_\psi \\ -s_\theta & s_\phi c_\theta & c_\phi c_\theta \end{pmatrix} \end{aligned} \tag{2.4} $$

where $c_x = \cos x$, $s_x = \sin x$, and $\mathrm{R}_x(\phi)$, $\mathrm{R}_y(\theta)$ and $\mathrm{R}_z(\psi)$ are the canonical rotations.

While intuitively and algebraically straightforward, the main drawback of this representation is that the Euler angles are not unique for a given orientation. This is due to both their modulo nature (2π and 0 represent the same angle) and the fact that some orientations can be represented by entirely different Euler angles (for example, $R(0,\pi,0) \equiv R(\pi,0,\pi)$). Unlike Euler angles, normalized quaternions offer a unique representation for all orientations[1] (see [36, 102] for a detailed discussion). Quaternion $\mathbf{q}$ is written as the 4-vector

$$\mathbf{q} = (S, \mathbf{U}) = (S, U_x, U_y, U_z) = S + \mathrm{i}U_x + \mathrm{j}U_y + \mathrm{k}U_z \tag{2.5}$$

A normalized quaternion satisfies the constraint $S^2 + U_x^2 + U_y^2 + U_z^2 = 1$. Quaternions are an extension of complex numbers, where S is the real component and U_x, U_y and U_z are coordinates along orthogonal imaginary axes represented by i, j and k, where $\mathrm{i}^2 = \mathrm{j}^2 = \mathrm{k}^2 = -1$. The connection between a quaternion $\mathbf{q}$ and a rotation of angle α about an axis represented by the inhomogeneous vector $\mathbf{A}$ is

$$\mathbf{q}(\alpha, \mathbf{A}) = (\cos(\alpha/2), \mathbf{A}\sin(\alpha/2)) \tag{2.6}$$

It is instructive here to note the similarity between equation (2.6) and the relationship between Cartesian and polar representations of complex numbers, that is $\mathbf{c}(r,\theta) = r\cos(\theta) + ir\sin(\theta)$. We can also write the relationship between Euler angles ϕ, θ, ψ and the equivalent quaternion as:

$$S = \cos\tfrac{1}{2}\phi\cos\tfrac{1}{2}\theta\cos\tfrac{1}{2}\psi + \sin\tfrac{1}{2}\phi\sin\tfrac{1}{2}\theta\sin\tfrac{1}{2}\psi \tag{2.7}$$
$$U_x = \sin\tfrac{1}{2}\phi\cos\tfrac{1}{2}\theta\cos\tfrac{1}{2}\psi - \cos\tfrac{1}{2}\phi\sin\tfrac{1}{2}\theta\sin\tfrac{1}{2}\psi \tag{2.8}$$
$$U_y = \cos\tfrac{1}{2}\phi\sin\tfrac{1}{2}\theta\cos\tfrac{1}{2}\psi + \sin\tfrac{1}{2}\phi\cos\tfrac{1}{2}\theta\sin\tfrac{1}{2}\psi \tag{2.9}$$
$$U_z = \cos\tfrac{1}{2}\phi\cos\tfrac{1}{2}\theta\sin\tfrac{1}{2}\psi - \sin\tfrac{1}{2}\phi\sin\tfrac{1}{2}\theta\cos\tfrac{1}{2}\psi \tag{2.10}$$

Quaternions possess a number of useful properties for manipulating rotations. A rotation in the opposite direction of a normalized quaternion is its complex conjugate:

$$\mathbf{q}(-\alpha, \mathbf{A}) = \mathbf{q}'(\alpha, \mathbf{A}) = (S, -\mathbf{U}) \tag{2.11}$$

Furthermore, like rotation matrices, quaternions can be multiplied to concatenate transformations; the final orientation $\mathbf{q}_3$ after rotating a frame by $\mathbf{q}_1$ and then $\mathbf{q}_2$ is given by the complex non-commutative[2] product

$$\mathbf{q}_3 = \mathbf{q}_2 * \mathbf{q}_1 = (s_2 s_1 - \mathbf{u}_2 \cdot \mathbf{u}_1, s_2\mathbf{u}_1 + s_1\mathbf{u}_2 + \mathbf{u}_2 \times \mathbf{u}_1) \tag{2.12}$$

were the $*$ operator represents complex multiplication, $\cdot$ is the vector dot product and $\times$ is the vector cross product[3]. Unfortunately, the redundant parameter and normalization constraint required by this representation make quaternions difficult to

[1] Strictly, there are two representations for a given orientation, one being obtained from the other by inverting both the angle and axis of rotation. This ambiguity is avoided by requiring the quaternion representation to have $S \geq 0$.

[2] Order is important when concatenating rotations in general, so $\mathbf{q}_1 * \mathbf{q}_2 \neq \mathbf{q}_2 * \mathbf{q}_1$ and $R_1R_1 \neq R_2R_1$

[3] Equation (2.12) can be obtained from equation (2.5) and the identities $\mathrm{ij} = -\mathrm{ji} = \mathrm{k}$, $\mathrm{jk} = -\mathrm{kj} = \mathrm{i}$ and $\mathrm{ki} = -\mathrm{ik} = \mathrm{j}$.

use in tracking filters. In practice, a combination of Euler angles and quaternions are employed to represent orientations (see Section 6.5).

For completion, the angle and axis of rotation corresponding to quaternion $\mathbf{q} = (s, \mathbf{u})$ are $\theta = 2\cos^{-1}(s)$ and $\mathbf{A} = \mathbf{u}^\top/|\mathbf{u}|$, and the corresponding Euler angles are:

$$\phi = \tan^{-1}(2(u_y u_z + s u_x)/(s^2 - u_x^2 - u_y^2 + u_z^2)) \tag{2.13}$$

$$\theta = \sin^{-1}(-2(u_x u_z - s u_y)) \tag{2.14}$$

$$\psi = \tan^{-1}(2(u_x u_y + s u_z)/(s^2 + u_x^2 - u_y^2 - u_z^2)) \tag{2.15}$$

All other transformations between Euler angle, axis/angle, quaternion and rotation matrix representations follow from combinations of the above transformations.

2.1.4 Pose and Velocity Screw

As described in Section 2.1.2, the pose of frame A with respect to frame B can be represented as the 4×4 homogeneous matrix ${}^B\mathrm{H}_A$. However, following the above discussion it is now clear that the *pose* (position and orientation) of a frame in $\mathbb{R}^3$ can be represented by only six degrees of freedom: three for position and three for orientation. Thus, the pose of a frame (or object) is often written more compactly as a *pose vector* $\mathbf{p} = (X, Y, Z, \phi, \theta, \psi)^\top$, where X, Y and Z represent position and ϕ, θ and ψ are the Euler angles describing orientation. Associated with this representation is the *velocity screw* $\dot{\mathbf{r}} = (\mathbf{V}, \boldsymbol{\Omega})^\top = (\dot{X}, \dot{Y}, \dot{Z}, \dot{\phi}, \dot{\theta}, \dot{\psi})^\top$, which encodes the time derivative of the pose vector parameters.

The pose of an object in frame A can be transformed to frame B by constructing the equivalent homogeneous matrix representation of the pose and multiplying by the coordinate transformation ${}^B\mathrm{H}_A$:

$$\mathrm{H}({}^B\mathbf{p}) = {}^B\mathrm{H}_A \mathrm{H}({}^A\mathbf{p}) \tag{2.16}$$

where $\mathrm{H}({}^A\mathbf{p})$ and $\mathrm{H}({}^B\mathbf{p})$ are the homogeneous matrix representation of pose vectors ${}^A\mathbf{p}$ and ${}^B\mathbf{p}$, constructed from equations (2.3)-(2.4). Alternatively, representing the position of an object in frame A as ${}^A\mathbf{T}$ and the orientation as ${}^A\mathrm{R}$, the transformation in equation (2.16) expands as

$${}^B\mathrm{R} = {}^B\mathrm{R}_A {}^A\mathrm{R} \tag{2.17}$$

$${}^B\mathbf{T} = {}^B\mathrm{R}_A {}^A\mathbf{T} + {}^B\mathbf{T}_A \tag{2.18}$$

where the relationship between ${}^B\mathbf{T}_A$, ${}^B\mathrm{R}_A$ and ${}^B\mathrm{H}_A$ is defined in equation (2.3). The velocity screw ${}^A\dot{\mathbf{r}} = ({}^A\mathbf{V}, {}^A\boldsymbol{\Omega})^\top$ of an object in frame A transforms to ${}^B\dot{\mathbf{r}} = ({}^B\mathbf{V}, {}^B\boldsymbol{\Omega})^\top$ in frame B as (see [68]):

$${}^B\boldsymbol{\Omega} = {}^B\mathrm{R}_A {}^A\boldsymbol{\Omega} \tag{2.19}$$

$${}^B\mathbf{V} = {}^B\mathrm{R}_A[{}^A\mathbf{V} - {}^A\boldsymbol{\Omega} \times {}^B\mathbf{T}_A] \tag{2.20}$$

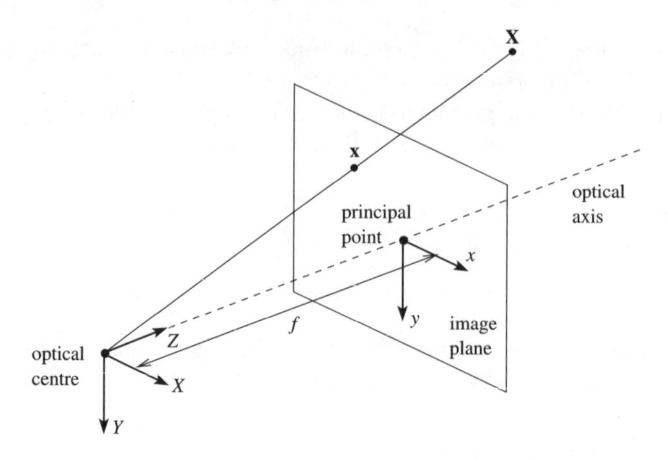

Fig. 2.2. Pin-hole camera model.

2.2 Sensor Models

This section describes how to analytically model the major components of a robotic vision system, namely the projective camera and active stereo head. Specific models for the experimental platform described in Chapter 7 are developed, but the general approach applies to a wide range of platforms. These models will be revisited in later chapters as the basis of algorithms and error analyses.

2.2.1 Pin-Hole Camera Model

A camera can be described as a system that performs the non-invertible mapping from real-space coordinates to image plane coordinates ($\mathbb{R}^3 \mapsto \mathbb{R}^2$). An analytical model for this transformation is essential for solving stereo reconstructions and predicting image plane measurements from an internal world model. The most commonly adopted camera model in computer vision literature is the *pin-hole* (or *central projection*) model. This model formally applies to a particular type of lensless camera, but in practice offers a good approximation to most lensed cameras. For cases in which the pin-hole approximation breaks down, such as zoom lens cameras, a more general *thick lens* model is required [94].

Figure 2.2 illustrates the projection of point **X** in real space onto the image plane at **x** for a simple pin-hole camera located at the origin of the world frame. This projection can be expressed as the non-linear transformation:

$$x = fX/Z \tag{2.21}$$

$$y = fY/Z \tag{2.22}$$

In the framework of projective geometry (see [53]), the operation of a camera can be described more compactly by a linear projective transformation:

$$\mathbf{x} = \mathrm{P}\mathbf{X} \tag{2.23}$$

where $\mathbf{X}$ and $\mathbf{x}$ are homogeneous coordinates in the world frame and image plane respectively, and P is the 3×4 *projection matrix* of the camera.

The inhomogeneous projection in (2.21)-(2.22) assumes that the camera is located at the origin of the world frame. When this is not the case, $\mathbf{X}$ must be transformed from the world frame to the camera frame before applying the projection. The position and orientation of the camera in the world frame are commonly called the *extrinsic camera parameters*. Representing these by an inhomogeneous translation vector $\mathbf{T}$ and rotation matrix R, a new projection matrix can be composed to perform both the coordinate transformation and projection onto the image plane:

$$\mathrm{P} = \mathrm{K}(\mathrm{R}^\top | - \mathrm{R}^\top \mathbf{T}) \tag{2.24}$$

where $(\mathrm{R}^\top | - \mathrm{R}^\top \mathbf{T})$ is a 3×4 matrix constructed from the extrinsic camera parameters (noting that $\mathrm{R}^{-1} \equiv \mathrm{R}^\top$ for a rotation matrix), and K is the 3×3 *camera calibration matrix*. The calibration matrix for a general pin-hole camera is

$$\mathrm{K} = \begin{pmatrix} af & sf & x_0 \\ 0 & f & y_0 \\ 0 & 0 & 1 \end{pmatrix} \tag{2.25}$$

where f is the focal length, x_0 and y_0 locate the intersection of the optical axis and image plane (also known as the principal point), a is the pixel aspect ratio and s describes pixel skew. These quantities are collectively known as the *intrinsic camera parameters*, since they are independent of the position and orientation of the camera. Recent advances in *camera self-calibration* have resulted in several methods to determine the calibration matrix automatically from sequences of images [124]. When accurate calibration is not required, it is usually quite reasonable to assume that a typical CCD or CMOS camera has square pixels (unity aspect ratio and zero skew) and a principal point at the origin of the image plane. This leads to a calibration matrix parameterized by a single intrinsic parameter, the focal length f:

$$\mathrm{K} = \begin{pmatrix} f & 0 & 0 \\ 0 & f & 0 \\ 0 & 0 & 1 \end{pmatrix} \tag{2.26}$$

The models presented in later chapters will adopt this approximation, with the focal length taken from known specifications, and the extrinsic parameters calibrated manually, as described in Appendix A.

2.2.2 Radial Distortion

Radial lens distortion is a deviation from the ideal pin-hole model described above and affects most lensed cameras. The characteristic *barrel-roll* effect is evident in the image of a grid shown in Figure 2.3(a), which was taken with a typical CCD

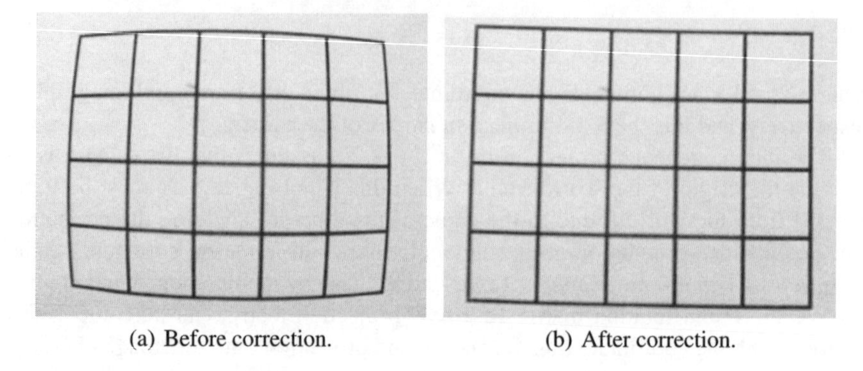

(a) Before correction. (b) After correction.

Fig. 2.3. Radial distortion and correction.

camera. This distortion would lead to significant errors in depth measurement and tracking if left uncorrected. To correct the image, the distortion is modelled as a radial displacement of image plane coordinates as follows:

$$\mathbf{x}_r = \mathbf{c}_r + (\mathbf{x} - \mathbf{c}_r) \left[1 + \sum_{i=1}^{\infty} K_i |\mathbf{x} - \mathbf{c}_r|^i \right] \tag{2.27}$$

where $\mathbf{x}$ is the position of an undistorted feature, $\mathbf{x}_r$ is the distorted position and $\mathbf{c}_r$ is the radial distortion centre (not necessarily at the optical centre). In practice, the distortion function is approximated by estimating only the first four coefficients $K_i, i = 1, \ldots, 4$.

It is a simple matter to estimate the parameters in equation 2.27 based on knowledge that the grid in Figure 2.3 should contain only straight lines when observed through an ideal pin-hole camera. The image in Figure 2.3(a) has been corrected using the following iterative algorithm based on this idea. At each iteration, the current estimate of the K_i's are used to warp the original image by replacing each pixel $\mathbf{x}$ with the pixel at the distorted position $\mathbf{x}_r$ calculated from equation (2.27). The warped image is then processed using edge extraction, centroid calculations and connectivity analysis to automatically extract a set of points on each grid-line. Straight lines fitted to each set of points using linear regression and the residual errors are tallied into a single cost associated with the current K_i. The simplex algorithm [112] is used to iteratively search over K_i to find the distortion parameters that lead to the minimal cost and result in grid-lines with minimum curvature, as shown in Figure 2.3(b). During normal operation, the correction is applied to every captured frame before further processing.

2.2.3 Active Stereo Camera Head

Many of the techniques in this book rely on both active vision and recovery of structure in three dimensions. Depth perception has been studied in computer vision in a

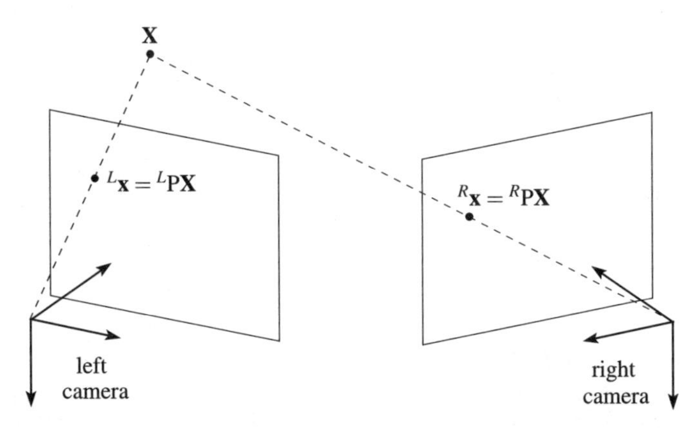

Fig. 2.4. Reconstruction of a point $\mathbf{X}$ from corresponding measurements $^L\mathbf{x}$ and $^R\mathbf{x}$ from stereo cameras with known projection matrices $^L\mathrm{P}$ and $^R\mathrm{P}$.

variety of forms for several decades. Depth can be recovered from a single image by exploiting cues such as focus, shading and texture [73] or utilizing a prior model of the object under observation (as discussed in Chapter 5). With two or more views of a scene, depth can be recovered by exploiting geometric constraints. *Structure from motion* has gained recent popularity and allows a scene to be reconstructed using many views from a single moving camera [77, 123]. However, the minimal and most commonly used multiple view method in robotics is *stereo vision*.

Figure 2.4 illustrates the well known principles of stereo reconstruction. Let $^L\mathbf{x}$ and $^R\mathbf{x}$ represent the projection of $\mathbf{X}$ onto the image planes of two cameras with different viewpoints. The stereo reconstruction of $\mathbf{X}$ is the intersection of the back-projected rays (shown as dotted lines) from the camera centres through $^L\mathbf{x}$ and $^R\mathbf{x}$, which can be found via the known projection matrices $^L\mathrm{P}$ and $^R\mathrm{P}$. In practice, stereo reconstruction presents two main difficulties. Firstly, corresponding measurements of each 3D point must be found on both image planes. This problem is simplified by considering the *epipolar geometry* established by the relative pose of the cameras, which constrains the possible locations of corresponding measurements. Secondly, if the extrinsic parameters of the cameras are not precisely known, the back-projected rays may not intersect anywhere. In this case, some optimal estimate of $\mathbf{X}$ is required [54]. Both problems are exacerbated by measurement noise.

Active vision refers to the ability to dynamically adjust the viewpoint of the cameras to aid in the selective acquisition and interpretation of visual cues[4]. Both stereo and active vision are important components of human perception, and an active stereo head is a mechanical device that aims to provide a robot with visuo-motor characteristics similar to a human head. Figure 2.5 illustrates a prototypical configu-

[4]Ambiguously, sensors that project energy into the environment are also referred to as *active*, and a light stripe sensor with this property is discussed in Chapter 3. However, the term "active vision" is generally reserved for dynamic viewpoint adjustment in this book.

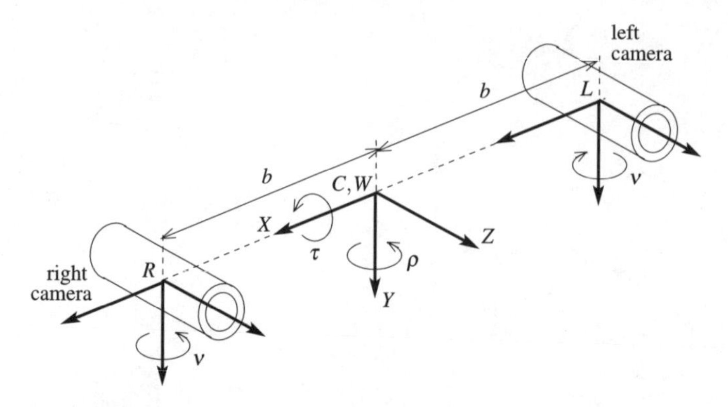

Fig. 2.5. Coordinate frames associated with the active vision platform.

ration of such a device, based on the "Biclops" head used experimentally in Chapter 7 (see Figure 7.1). Stereo cameras are separated by a baseline congruent with human vision, and rotational joints aligned to the y-axis of each camera enable the nominally parallel optical axes to converge or diverge. The Biclops verge mechanism constrains the cameras to rotate symmetrically in opposite directions, so vergence contributes only a single mechanical degree of freedom. A neck provides two additional rotational degrees of freedom to pan and tilt the cameras in unison. Beyond the minimalist Biclops design, active stereo heads have been constructed with several additional degrees of freedom (such as independent verge and roll), and multiple cameras for human-like peripheral and foveal visual fields [136].

The following angles, coordinate frames and transformations are defined to model the Biclops camera configuration shown in Figure 2.5:

- L and R are the left and right camera frames, separated by baseline $2b$ and centred at the focal points of the cameras. The direction of the X and Y-axes, parallel to the image plane of the cameras, are defined by the ordering of pixels on the image plane in computer memory: left and down. The Z-axes are parallel to the optical axes of the cameras, with the direction dictated by the right-handed coordinates. Angle ν denotes vergence, which is positive for L and negative for R when the cameras are converged.
- Stereo camera frame C is attached to the top of the neck between the cameras, and provides a reference for stereo reconstruction. Frames C, L and R move as the Biclops pans and tilts. The axes of C are parallel to L and R, and the latter are centred at

$$\mathbf{C}_{L,R} = (\mp b, 0, 0)^\top \tag{2.28}$$

with respect to C, taking the negative sign for L and the positive for R. Let ${}^{L}\mathrm{P}_C$ and ${}^{R}\mathrm{P}_C$ represent the projection matrices of the left and right cameras in C. To simplify later models, the cameras are assumed to have identical focal length. Furthermore, Section 2.2.4 will show that *projective rectification* allows us to treat the verge angle as zero, and set R = I in equation (2.24). Under these conditions, the left and right projection matrices in C are given by

$$ {}^{L,R}\mathrm{P}_C = \begin{pmatrix} f & 0 & 0 & \pm fb \\ 0 & f & 0 & 0 \\ 0 & 0 & 1 & 0 \end{pmatrix} \tag{2.29} $$

- World frame W is defined at the same position as C, but is attached to the base of the neck and remains stationary with respect to the body of the robot. Transforming reconstructed measurements from C to W compensates for ego-motion as the Biclops head pans and tilts. Representing the pan angle as ρ and the tilt angle as τ, the homogeneous transformation from C to W is

$$ {}^{W}\mathrm{H}_C(\rho,\tau) = \begin{pmatrix} \cos\rho & \sin\rho\sin\tau & 0 & 0 \\ 0 & \cos\tau & -\sin\tau & 0 \\ -\sin\rho & \cos\rho\sin\tau & \cos\rho\cos\tau & 0 \\ 0 & 0 & 0 & 1 \end{pmatrix} \tag{2.30} $$

A simple active vision strategy commonly encountered in the literature is to maintain a moving feature (say, a tracked target) near the centre of the image plane so that it always remains visible. Let ${}^{L}\mathbf{x} = ({}^{L}x, {}^{L}y)^{\top}$ and ${}^{R}\mathbf{x} = ({}^{R}x, {}^{R}y)^{\top}$ represent stereo measurements of the tracked feature, and $\dot{\omega}_{\rho}$, $\dot{\omega}_{\tau}$ and $\dot{\omega}_{v}$ represent the angular pan, tilt and verge velocities required to keep the target centred. A simple proportional control law to calculate these velocities is

$$ \dot{\omega}_{\rho} = k_{\rho}({}^{L}x + {}^{R}x) \tag{2.31} $$
$$ \dot{\omega}_{\tau} = k_{\tau}({}^{L}y + {}^{R}y) \tag{2.32} $$
$$ \dot{\omega}_{v} = k_{v}({}^{L}x - {}^{R}x) \tag{2.33} $$

with suitably chosen gains k_{ρ}, k_{τ} and k_{v}. This particular form of active vision is known as *smooth pursuit tracking*. An alternative form of active tracking is *saccading*, which involves short, rapid motions to centre the target.

2.2.4 Rectilinear Stereo and Projective Rectification

One particular configuration of stereo cameras deserves special consideration: the alignment of parallel camera axes and coplanar image planes, also known as *rectilinear stereo*. The simple camera projection matrices that arise in this case were given in equation (2.29). The usefulness of rectilinear stereo arises from a simple epipolar geometry that constrains corresponding stereo measurements to lie on the same scanline in both images, facilitating the search for stereo correspondences. Furthermore, the reconstruction equations have a simple form for rectilinear stereo. A particularly useful property of rectilinear stereo is that, for a general non-rectilinear configuration, we can always recover equivalent rectilinear stereo measurements by applying a transformation known as *projective rectification*. In general, the transformation to rectify stereo images can be determined purely from a set of point correspondences (see [53]). However, since the Biclops configuration is known we can calculate the transformation directly from the measured verge angle.

The following formulation is based on the knowledge that when a camera undergoes a fixed rotation, the image plane coordinates undergo a known projective transformation. Thus, the equivalent image prior to the rotation can be recovered by applying the inverse transformation. Let the verge angle ν represent the rotation about the y-axis between the rectilinear frame and the rotated camera frame, and let P represent the projection matrix of a camera centred at the origin. Also, let $\mathbf{x}_m$ represent the measured location of a feature in the unrectified image, and $\mathbf{x}_r$ represent the equivalent feature after rectification. A real-space point $\mathbf{X}_r$ projecting onto $\mathbf{x}_r$ can be recovered (up to an unknown scale) as

$$\mathbf{X}_r = \mathrm{P}^+ \mathbf{x}_r \tag{2.34}$$

where P^+ is the pseudo inverse of P, given by

$$\mathrm{P}^+ = \mathrm{P}^\top (\mathrm{P}\mathrm{P}^\top)^{-1} \tag{2.35}$$

such that $\mathrm{P}\mathrm{P}^+\mathbf{x} = \mathbf{x}$ [52]. For a measurement at $\mathbf{x}_r = (x_r, y_r, 1)^\top$, equation (2.34) gives $\mathbf{X}_r = (x/f, y/f, 1, 0)^\top$, which is the point at infinity in the direction of the ray back-projected through $\mathbf{x}_r$. Now, let $\mathrm{R}_y(\nu)$ represent the rotation that transforms points from the rectilinear frame to the current camera frame. By rotating $\mathbf{X}_r$ and re-projecting into the image plane, the measurement corresponding to the rectified feature is calculated as

$$\mathbf{x}_m = \mathrm{P}\mathrm{R}_y(\nu)\mathrm{P}^+\mathbf{x}_r \tag{2.36}$$

For the pin-hole camera model given by equations (2.21)-(2.22), the transformation above can be expressed in inhomogeneous coordinates as

$$x_m = \frac{(x_r \cos\nu + f \sin\nu) f}{f \cos\nu - x_r \sin\nu} \tag{2.37}$$

$$y_m = \frac{y_r f}{f \cos\nu - x_r \sin\nu} \tag{2.38}$$

To apply projective rectification, the position of each pixel in the rectified image is transformed by equations (2.37)-(2.38) to locate the corresponding pixel in the measured image. Since the left and right cameras verge in opposite directions, the above transformation is applied with angle $-\nu$ for L and $+\nu$ for R. The transformation can be implemented efficiently as a lookup table of pixel coordinates since it remains fixed from frame to frame unless the verge angle changes. For the experimental work presented in later chapters, rectification is applied to all captured images to simplify subsequent operations.

2.3 From Images to Perception

2.3.1 Digital Image Processing

A digital image is a quantized, discrete 2-dimensional signal formed by sampling the intensity and frequency of incident light on the image plane. Typically, each sample is formed by integrating light for a short period over a small area on a 2D sensor array such as a *charged coupled device* (CCD) or CMOS transistor array. Analytically, we represent a digital image as a $W \times H$ array $\mathrm{I}(x,y)$, with $x = 0,1,..,W-1$ and $y = 0,1,..H-1$ spanning the width and height of the image. The term *pixel* describes a single element of the array, while *scan-line* refers to an entire row. In practice, pixels in the sensor array are addressed with $I(0,0)$ at the top left, x increasing to the right and y increasing downward, which accounts for the choice of frame orientation in Figures 2.2 and 2.5. To represent colour, pixels may be vector-valued with components representing the intensity of incident light in red, green and blue frequency *bands* (or *channels*). Image processing also frequently involves *grayscale* images in which pixels represent overall intensity, and *binary* images that encode the presence or absence of features (such as a particular colour) as boolean-valued pixels.

Along the path from images to perception, the progression through low-level, mid-level and high-level vision provides an illustrative (though simplistic) road map of the archetypal computer vision algorithm. Operations at successive levels are distinguished by their increasing abstraction away from the raw images, and decreasing awareness of incidental factors such as lighting and viewpoint. The first and lowest level of processing, also known as early vision, typically involves detecting features such as colours or spatial and temporal intensity gradients, and matching features for depth recovery and optical flow. Mid-level vision is concerned with forming low-level features into coherent entities through segmentation, model-fitting and tracking. Finally, high-level algorithms distill usable semantics by classifying or recognizing objects and behavioural patterns. Many excellent texts cover all levels of the computer vision in greater detail, and the interested read is referred to [36, 39, 52].

2.3.2 The Role of Abstraction

One of the fundamental issues in perception for robotic systems is the choice of representation, that is, how the environment and goals are abstracted. Traditional robotic systems employ a hierarchy of *sensing*, *planning* and *action* modules, with suitable abstractions of the world providing the "glue". The role of perception is then to reduce the real world to a set of relevant semantic tokens (such as recognized objects), to form an abstract internal model for high-level processing.

An alternative approach proposed by Brooks [16] is *intelligence without representation* in which the notion of a centralized world model is abandoned. Brooks proposes that instead of the traditional functional components, control should be implemented as a layered set of competing *activities*. Each activity independently couples perception to action with little or no intermediate abstraction, and careful design of the interactions between layers leads to emergent intelligence. This approach facilitates simple and robust perception algorithms. Many robotic grasping methods found in the literature can be used in this spirit. Rodrigues *et al.* [130] describe an algorithm for finding three-fingered grasp points on the boundary of an arbitrary 2D object directly from a binary image. Semantic-free grasp planning algorithms based on moments of inertia and boundary tracing, using 2D [40, 55] or 3D visual data [46], have also been demonstrated. Image-based visual servoing (discussed in detail below) controls the pose of a robot using low-level visual measurements such as corners and lines, without further abstraction. Combining these algorithms would enable a service robot to grasp arbitrary, visually sensed targets without imposing any interpretations on the measurements.

While many low-level behaviours can be performed directly on measurements, abstractions can be very useful at higher levels of planning. For example, classifying an object as a telephone allows a robot to minimize the grasp planning problem by exploiting prior knowledge about telephones. Furthermore, semantics extracted from available measurements combined with prior knowledge (whether programmed or learned) can be used to infer structure that may be hidden from view.

Ultimately, abstractions are needed at some level to enable a robot to both learn and interact naturally with humans. Consider a task given as the high-level command: "Please fetch a bottle of water". The robot must sufficiently abstract the world to identify, at a minimum, objects satisfying the description of *bottle*. Thus, the robot must be programmed with, or preferably learn autonomously, a representation for bottles associated with this semantic label. A common approach to object recognition in robotics is to form a database of known objects from labelled training samples. Object models may be geometrical, view-based or use a combination of representations, and are created with varying degrees of human interaction. Many industrial service tasks involve only a small set of well defined objects, in which case predefined CAD models (typically edge-based) can be used to locate and track objects during manipulation. Examples of this approach include [41, 142, 152].

Other representations have been explored to deal with the wide variety of objects encountered in domestic applications, and to facilitate autonomous learning. The humanoid service robot developed by Kefalea *et al.* [80] represents objects using a set of sampled edge-based templates, each with an associated orientation and grasp configuration. The robot generates templates semi-autonomously by placing the object on a turntable. Given an unknown scene, all templates for every known object are elastically matched to the captured image to recover the identity, approximate pose and grasp information. Kragić [88] describes a similar scheme using a set of image templates and associated poses in a principal components analysis framework. In this case, each object is also represented by an edge-based geometric model used to refine the initial pose and track the object during grasping. In related work, Ekvall *et al.* [33] replace the image templates with colour cooccurrence histograms, which represent the spatial distribution of colour with scale and rotation independence. Objects are identified using a histogram similarity measure, and the pose is recovered as a weighted mean of the orientation associated with different views. A variety of other representations for object recognition and pose estimation for robotic manipulation can be found in [85, 113, 146].

The approaches described above require an explicit model for each instance of an object. However, a kitchen may contain dozens of cups with unique variations in colour, texture and geometry that require each to be modelled separately. As the number of objects in the database increases so does the burden on storage space and the computational expense of matching. One solution to this problem is to represent classes, such as "cups", rather than specific objects. Much recent work has focussed on the similar problem of detecting faces or people in images independently of the individual (for example, see [160]). For classifying domestic objects, data-driven geometric modelling is a popular solution. Geometric modelling has been demonstrated using bivariate polynomials [12], Generalized Cylinders [23, 126] and primitives such as planes, spheres and cones [100, 167]. Once a scene has been modelled, objects such as cups can identified as specific collections of primitives without storing a model for each cup. Data-driven modelling and classification will be explored further in Chapter 4.

2.4 From Perception to Control

The principal control task in traditional industrial robotics is to accurately, rapidly and repeatably drive the end-effector through a sequence of pre-programmed poses. Perception plays little part since both the robot and environment are completely defined. In contrast, service robots may only need to match the speed and accuracy of their human partners, while service tasks may be poorly defined and require little repetition. This leads to control algorithms with a distinctly different aim: to robustly close the loop between perception and action. The major approaches to visually guided robotic grasping are discussed in the following sections.

The classical framework of *look-then-move* or open-loop kinematic control solves the important problem of uncertain target location. Typical implementations of this approach are used to control the humanoid robots in [10, 104]. The target is localized using calibrated cameras and the required joint angles for grasping are calculated using inverse kinematics. The obvious drawback of this end-effector open-loop approach is that kinematic and camera calibration errors can still defeat the robot.

Calibration errors can be minimized by directly observing the gripper in addition to the target, and two distinct approaches to this problem have been demonstrated. The first is to construct a mapping between robot joint angles and locations in camera space by observing the gripper at various locations. Given a target pose, this mapping then provides the required joint angles. This form of open-loop control is exemplified by the learning-based schemes discussed in Section 2.4.1. The second approach, known as *visual servoing*, is to continuously measure the pose of both the gripper and target during grasping, and drive the joint angles as a function of the pose error. Naturally, the gripper will be driven toward the target until the pose error reduces to zero. The various visual servoing schemes are discussed further in Section 2.4.2.

It is important to recognize that end-effector open and closed-loop schemes offer complementary advantages. Closed-loop control achieves low sensitivity to calibration errors but constrains the gripper to be observable, while open-loop control is less accurate with fewer constraints. Gaskett and Cheng [43] attempt to exploit the advantages of both approaches by combining learning-based control with conventional visual servoing. The learned mapping between camera space and joint angles provides a gross motion to bring the gripper within view, after which the closed-loop controller takes over. Chapter 6 describes an alternative framework for combining both open and closed-loop visual control.

2.4.1 Learning-Based Methods

Learning algorithms are usually inspired by advances in our understanding of human learning from studies in neuroscience. One such idea that has found popularity in robotics is the concept of the *visuo-motor mapping*, which describes the relationship between visually perceived features and the motor signals necessary to act upon them. One of the elementary problems in visuo-motor mapping is learning to *saccade*; generating the eye motion required to center a visual stimulus at the fovea. In [127], the saccadic visuo-motor map for an active camera head is represented as a vector of motor signals at each image location, learned via an iterative process involving random saccades. Learning-based control of an active head has also been used for the related problem of *smooth pursuit* tracking [42].

Another visuo-motor mapping can be defined between cameras and a robot arm. Ritter *et al.* [128] apply this idea to controlling a three-link robotic arm using stereo cameras. The mapping is implemented as a neural network that takes as input a target location specified as corresponding points on the stereo image planes, and output the three joint angles required to drive the arm to that location. The network is trained by driving the arm to random locations and observing the gripper. The humanoid robot described by Marjanović *et al.* [99] autonomously learns a visuo-motor mapping to point at visual targets through trial and error, with no knowledge of kinematics. Rather than mapping joint angles directly, pointing motions are constructed from a basis of predefined arm positions. The visuo-motor mapping generates weights to linearly interpolate these base poses and drive the arm in the desired direction.

Learning-based methods are also exploited at the task planning level, and are capable of addressing complex issues. For example, the system described by Wheeler *et al.* [162] learns how to grasp objects to satisfy future constraints, such as how the object will be placed (known as *prospective planning*). *Programming by demonstration*[5] is a popular planning framework that has been demonstrated both for industrial [70] and service robots [32, 86] (see [137] for a comprehensive review). In this approach, the robot observes the actions and subsequent environmental changes made by a human teacher and stores the sequence for later repetition. Programming by demonstration eases the problem of planning complex manipulations, but raises a number of other issues: differing kinematic constraints of the teacher and robot require adaption of observations [7]; the robot must be able to discriminate task-oriented motions from incidental motions and measurement noise [21]; suitable abstractions must be constructed so that learned motions can be applied in different circumstances [1].

2.4.2 Visual Servoing

Visual servoing describes a class of closed-loop feedback control algorithms in robotics for which the control error is defined in terms of visual measurements. Like all

[5] also known in the literature as *learning from demonstration*, *imitation learning* and *assembly-plan-from-observation*.

feedback systems, the advantage of this approach is to minimize the effect of disturbances and uncertainty in kinematic and sensor models. The effectiveness of the visual servoing paradigm is demonstrated by an extensive literature and diverse applications, including industrial automation [75], assistance in medical operations [91], control of ground vehicles [158], airborne vehicles [159], and robotic grasping. A comprehensive tutorial and review of visual servoing for robotic manipulators can be found in [68], and a review of more recent developments can also be found in [89].

The visual servoing literature has established a well-defined taxonomy of popular configurations and control algorithms. At the kinematic level, the relationship between the robot and camera is often described as *eye-in-hand* or *fixed-camera*. In the former configuration, the camera is rigidly attached to the end-effector and the goal is to control the view of the camera (and thus the pose of the gripper) with respect to a target. Conversely, a fixed-camera configuration places the camera and end-effector at opposing ends of a kinematic chain, and pose of the gripper is controlled independently of the camera view. This latter configuration is more akin to human hand-eye coordination.

A systematic categorization of visual servoing control architectures was first introduced by Sanderson and Weiss [135]. One of the principal conceptual divisions is the distinction between *position-based* and *image-based* visual servoing, as illustrated in Figure 2.6 for a fixed-camera configuration. In position-based visual servoing, the control error is calculated after reconstructing the 6D pose of the gripper from visual measurements, typically via 3D model-fitting. Joint angles are then driven by the error between the observed and desired pose. Conversely, the control error in image-based visual servoing is formulated directly as the difference between the observed and desired location of features (such as points and edges) on the image plane. It is interesting to note that a recent comparison of position-based and image-based servoing concluded that both exhibit similar stability, robustness and sensitivity to calibration errors [31].

Visual servoing schemes can be further classified as *direct visual servo*[6] or *dynamic look-and-move*, depending on whether the control law directly generates joint motions, or Cartesian set points for lower-level joint controllers. This distinction is usually not made in the literature as almost all practical systems are regarded as dynamic look-and-move.

A final significant distinction can be made between *endpoint open-loop* (EOL) and *endpoint closed-loop* (ECL) control [68]. An ECL controller observes both the end-effector and target to determine the control error, while an EOL controller observes only the target. In the latter case, the relative position of the robot is controlled using a known, fixed camera to robot (*hand-eye*) transformation. The required hand-eye calibration is often based on observation of a calibration rig [156] or the motion of the end-effector [6], and has a significant effect on the accuracy of EOL control. Importantly, hand-eye calibration is not required for ECL control since the pose of the end-effector is measured directly. Positioning accuracy is therefore independent of kinematic uncertainty (although stability and other dynamic issues may arise).

---

[6] This terminology was introduced by Hutchinson *et al.* [68].

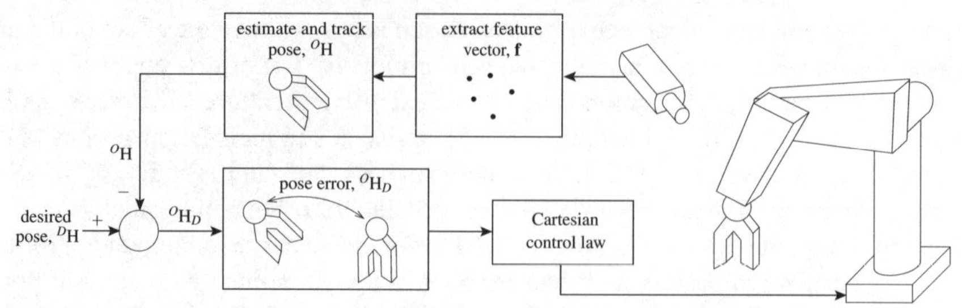

(a) Position-based visual servoing framework. The control error is the transformation $^{O}H_{D}$ between the desired pose ^{D}H and the observed pose ^{O}H. The observed pose is estimated from the visual feature vector, $\mathbf{f}$.

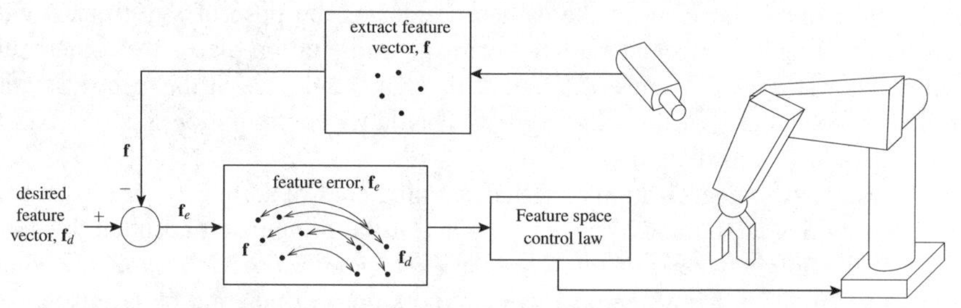

(b) Image-based visual servoing framework. The control error $\mathbf{f}_e$ is the difference between the observed image feature vector $\mathbf{f}$ and the desired feature vector $\mathbf{f}_d$.

Fig. 2.6. Comparison of position-based and image-based visual servoing.

Image-based visual servoing, which generally receives greater attention in the literature, has been studied for both eye-in-hand [34, 81, 115] and fixed-camera [63, 65] configurations. Typically, the desired locations of target features are obtained from a real image of the robot already in the desired pose. The robot can then be servoed back to the desired pose with a high degree of robustness to calibration errors and without knowledge of the 3D structure of the target. A central concept in image-based visual servoing is the *image Jacobian* (or *interaction matrix*), which relates differential changes in the 6D pose or joint angles of the end-effector to changes in the image-space location of features. The image Jacobian thus allows an image-space control error to be transformed into a pose or joint-space error. The image Jacobians for a variety of features, including points and lines, are derived in [34].

A well known drawback of classical image-based control is the existence of singularities in the image Jacobian and local minima in the control error, leading to instabilities and non-convergence [20]. Furthermore, calculation of the image Jacobian requires knowledge of depth, which may be recovered using 3D model-based tracking [87] or some approximation [97]. Alternatively, Hosada and Asada [65] dispose of analytical Jacobians altogether and avoid these problems by maintaining an on-line discrete-time estimate of the image Jacobian. A final significant drawback is that classical image-based visual servoing provides no explicit control of pose, which can lead to inefficient and unpredictable trajectories in Cartesian space. Recent results show that this issue can be avoided by decoupling the control of orientation and translation [83, 97, 129].

As noted earlier, position-based visual servoing requires explicit reconstruction of the robot and target pose, which imposes a greater computational burden than image-based control [68]. Furthermore, pose estimation typically relies on knowledge of the camera parameters and a precise 3D model of the object. In contrast to image-based control, the accuracy of the controller is therefore sensitive to both camera calibration and the chosen pose estimation algorithm [31]. On the other hand, controlling the pose of the end-effector in real space leads to predictable trajectories and allows simple path planners to be directly incorporated into the controller. However, since position-based control imposes no constraints on image-space trajectories, the observability of the target within the camera frame is no longer guaranteed. While this is considered one of the classical drawbacks of position-based control, Cervera and Martinet [18] demonstrated that simply formulating the control error to nullify the target pose leads to simple image plane trajectories that are likely to maintain the visibility of the target.

Wilson *et al.* [163] present the design and implementation of a classical EOL position-based visual servo controller. Using an end-effector mounted camera, the robot is able to maintain a desired pose relative to a target work-piece using a Kalman filter tracker based on point features. By achieving a sampling rate of 61 Hz in the visual control loop, this work demonstrates that pose estimation is not necessarily a computational burden. Wunsch and Hirzinger [165] present a similar real-time position-based scheme that tracks edge features instead of points. Tracking filters can simplify measurement association in model-based pose estimation, and therefore play a prominent role in position-based visual servoing. However, a drawback is the requirement of an explicit dynamic model for target pose prediction, since targets with poorly defined dynamics can easily lead to tracking failures. Adaptive Kalman filtering promises to alleviate this problem via a dynamically estimated motion model [164].

Various techniques have been proposed in the literature to overcome the difficulties of both image and position-based schemes. The *hybrid* approach of 2-1/2-D visual servoing [97] employs partial pose reconstruction to control the orientation of the end-effector in real space, while translation is controlled using *extended* image coordinates. A rough estimate of depth is required for translation control, but does not affect the positioning accuracy of the controller. Unlike conventional position-based control, 2-1/2-D servoing does not require precise hand-eye and camera calibration or a 3D model of the target. Furthermore, the controller produces linear trajectories in real space, while the singularities associated with the classical image Jacobian are avoided.

Other schemes aim to eliminate the need for accurate camera and kinematic calibration by employing linear approximations. For example, Cipolla and Hollinghurst [24] present an approach based on *affine stereo*. The hand-eye transformation and camera model are reduced to 16 linear coefficients that are robustly estimated from the stereo measurements of four known points. Visual servoing based on affine stereo is shown to be robust to linear kinematic errors, camera disturbances and perspective effects. Namba and Maru [110] achieve an approximate linear stereo reconstruction by parameterizing camera space using angular quantities, and also describe a linear approximation to the inverse kinematics of a humanoid robot. In this scheme, joint velocities are calculated directly as a linear function of the image plane error. Interestingly, it is straightforward to show that image and position-based control are equivalent for point-to-point alignment in a linear servoing framework [61]. However, as with any approximate method, the performance of this class of controllers is bounded by the validity of the linear approximation.

2.4.3 Summary

Sensor models play an important role in 3D model-based vision, which is a central theme of this book. The analytical tools commonly used in modelling robotic and vision systems were presented in Section 2.1, and Section 2.2 applied this framework to modelling the major components of a robotic vision system. In particular, the pinhole approximation of a lensed camera and the kinematic model of an active stereo head were described, which form the basis of several algorithms developed in later chapters. For example, the validation and reconstruction algorithm for robust light stripe sensing in Chapter 3, the 3D model-based tracker in Chapter 5 and the hybrid position-based visual servoing method in Chapter 6 all draw heavily on both camera and kinematic models.

The broad overview of perception and control techniques for robotic manipulation presented in Sections 2.3 and 2.4 serve primarily to show that much research remains to be done in this area. The adoption of 3D model-based vision in this book is motivated by several of the issues raised here. In particular, the chosen world representation defines the robot's ability to classify, autonomously learn and attach semantic meaning to unknown objects, which is important when working in an domestic environment. Object modelling and recognition using compositions of simple geometric primitives (autonomously extracted from range data) is a well known approach to addressing these issues, and is discussed further in Chapter 4.

Visual control algorithms are typically based on either biologically inspired learning or some form of visual servoing. The latter approach is pursued in this book as it provides greater robustness to the dynamic disturbances and modelling uncertainties likely encountered by a service robot. Fitting in with the 3D model-based vision paradigm, a position-based visual servoing algorithm based on the principles outlined above is presented in Chapter 6.

3

Shape Recovery Using Robust Light Stripe Scanning

Recovering the three dimensional structure of a scene accurately and robustly is important for object modelling and robotic grasp planning, which in turn are essential prerequisites for grasping unknown objects in a cluttered environment. Shape recovery techniques are broadly described as either *passive* or *active*. Passive methods include recovering shape from a single image using cues such as shading, texture or focus, and shape from multiple views using stereopsis or structure-from-motion. Passive shape recovery has relatively low power requirements, is non-destructive and more akin to our biological sensing modalities. However, the accuracy and reliability of passive techniques is critically dependent on the presence of sufficient image features and the absence of distractions such as reflections. Active sensing is necessary to achieve the accuracy required for the object modelling and tracking techniques described in the following chapters.

Active sensing relies on projecting energy into the environment and interpreting the modified sensory view. Active techniques generally achieve high accuracy and reliability, since the received signal is well-defined but at the cost of high power consumption. Active shape recovery is generally based on either some form of structured light projection or time-of-flight ranging. Recent developments in active CMOS time-of-flight cameras enable a complete 3D image to be captured in a single measurement cycle [93], however this technology is still limited by timing resolution. Structured light ranging is based on the same principles as stereopsis, but with a masked light pattern projected onto the scene to provide easily identifiable image features. The main research in this area has focused on designing coded light patterns to uniquely identify each surface point [106]. For robotic applications, the disadvantage of this approach is the requirement of a large, high-power pattern projector.

Light stripe ranging is a simple structured light technique that uses a light plane (typically generated using a laser diode) to reconstruct a single range profile of the target with each frame. This technique offers the advantages of compactness, low power consumption and minimal computational expense. However, applying light stripe sensing to service robots presents unique challenges. Conventional scanners require the brightness of the stripe to exceed that of other features in the image to be reliably detected, making the technique most effective in highly controlled environments. Robust light stripe detection methods have been proposed in previous work but suffer from issues including assumed scene structure, lack of error recovery and acquisition delay. Thus, the goal in this chapter is to develop a robust light stripe sensor for service robots that overcomes this major shortcoming.

The solution is an actively calibrated stereoscopic light stripe scanner, capable of robustly detecting the stripe in the presence of secondary reflections, cross-talk and other noise mechanisms. Measurement validation and noise rejection are achieved by exploiting the redundancy in stereo measurements (conventional scanners use a single camera). The validation and reconstruction algorithms are optimal with respect to sensor noise, which results in higher precision ranging than conventional methods. Furthermore, self-calibration from an arbitrary non-planar target allows robust validation to be achieved independently of ranging accuracy. Finally, operating in normal light allows colour and range measurements to be captured and implicitly registered in the same camera. Registered colour/range data is highly desirable for the object modelling and tracking techniques developed in the following chapters.

The following section introduces the basic principles of conventional single-camera light stripe range sensing and discusses existing robust techniques. The theoretical framework of the method proposed in this chapter is presented in Section 3.2. Section 3.3 details the active calibration process that allows the scanner to be calibrated from measurements of an arbitrary non-planar target. The experimental implementation of the scanner is described in Section 3.4, including low-level image processing for stripe detection and range data post-processing. Finally, Section 3.5 experimentally validates the proposed technique, including a comparison of the noise robustness achieved by the proposed scanner and other methods. VRML models of experimental colour/range scans are provided in the *Multimedia Extensions*, along with source code and data sets to implement the calibration and scanning algorithms.

3.1 Conventional Light Stripe Ranging and Related Work

Light stripe ranging is an active, triangulation-based technique for non-contact surface measurement that has been studied for several decades [3, 139]. A review of conventional light stripe scanning and related range sensing methods can be found in [9,12,59]. Range sensing is an important component of many robotic applications, and light stripe ranging has been applied to a variety of robotic tasks including navigation [4, 116], obstacle detection [57], object recognition for grasping [5, 126] and visual servoing [82].

Figure 3.1 illustrates the operation of a conventional single-camera light stripe sensor. The principle is similar to binocular stereo (see Section 2.2.3), with one of the cameras replaced by a light plane projector (typically a laser with a suitable lens). The stripe reflected from the target is measured by the camera and each point in the 3D surface profile (for example **X** in Figure 3.1) is reconstructed by triangulation, using the known transformation between the camera and projector. To capture

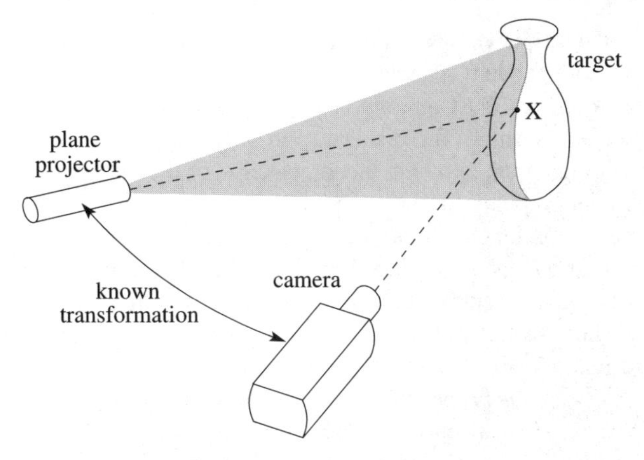

Fig. 3.1. Conventional single-camera light stripe sensor.

a complete range image, the light plane is mechanically panned across the target and the range slices are registered into a mesh[1].

The drawback of conventional single-camera light stripe ranging is that favourable lighting conditions and surface reflectance properties are required so the stripe can be identified as the brightest feature in the image. In practice, this is achieved by coating the target with a matte finish, using high contrast cameras or reducing the level of ambient light. When the range sensor is intended for use by a service robot to recognize unknown objects in a domestic or office environment [147], various noise mechanisms may interfere to defeat stripe detection: smooth surfaces cause secondary reflections, edges and textures may have a stripe-like appearance, and cross-talk can arise when multiple robots scan the same environment.

A number of techniques for improving the robustness of light stripe scanners have been proposed in other work, using both stereo and single-camera configurations. Magee *et al.* [96] develop a scanner for industrial inspection using stereo measurements of a single stripe. Spurious reflections are eliminated by combining stereo fields via a minimum intensity operation. This technique depends heavily on user intervention and *a priori* knowledge of the scanned target. Trucco *et al.* [155] also use stereo cameras to measure a laser stripe, and treat the system as two independent single-camera sensors. Robustness is achieved by imposing a number of consistency checks to validate the range data, the most significant of which requires independent single-camera reconstructions to agree within a threshold distance. Another constraint requires valid scan-lines to contain only a single stripe candidate, but a method for error recovery in the case of multiple candidates is not proposed. Thus, secondary reflections cause both the true and noisy measurements to be rejected.

[1] An alternative to rotating the light stripe is to rotate the target itself. For robotic applications, this could be achieved by grasping and rotating the object in the gripper. However, this approach is difficult in practice when 3D sensing is a prerequisite to grasp planning!

Nakano *et al.* [109] develop a similar method to reject false data by requiring consensus between independent scanners, but using two laser stripes and only a single camera. In addition to robust ranging, this configuration provides direct measurement of the surface normal. The disadvantage of this approach is that each image only recovers a single range point at the intersection of the two stripes, resulting in a significant acquisition delay for the complete image.

Other robust scanning techniques have been proposed using single-camera, single-stripe configurations. Nygards and Wernersson [117] identify specular reflections by moving the scanner relative to the scene and analyzing the motion of reconstructed range data. In [57], periodic intensity modulation distinguishes the stripe from random noise. Both of these methods require data to be associated between multiple images, which is prone to error. Furthermore, intensity modulation does not disambiguate secondary reflections, which vary in unison with the true stripe. Alternatively, Clark *et al.* [25] use linearly polarized light to reject secondary reflections from metallic surfaces, based on the observation that polarized light changes phase with each specular reflection. However, the complicated acquisition process requires multiple measurements through different polarizing filters.

Unlike the above robust techniques, the method described below uniformly rejects interference due to secondary reflections, cross-talk, background features and other noise mechanisms. A significant improvement over previous techniques for error detection is a mechanism for the *recovery* of valid measurements from a set of noisy candidates. The reconstructed depth data is optimal with respect to sensor noise, unlike the stereo techniques in [109, 155], and stereo measurements are fused with the light plane parameters to provide greater precision than a single-camera configuration. Finally, conventional techniques using a special camera for stripe detection typically require a second camera to measure colour [58]. The ability to operate in ambient indoor light allows our sensor to measure implicitly registered colour and range in the same camera.

3.2 Robust Stereoscopic Light Stripe Sensing

As discussed above, the shortcomings of light stripe sensors arise from the difficulty in disambiguating the primary reflection of the stripe from secondary reflections, cross-talk and other sources of noise. The following sections detail the principles of an optimal strategy to resolve this ambiguity and robustly identify the true stripe by exploiting the redundancy in stereo measurements.

3.2.1 Problem Statement

Figure 3.2 shows a simplified plan view of a stereoscopic light stripe scanner to demonstrate the issues involved in robust stripe detection. The light plane projector is located between a pair of stereo cameras, which are arranged in a rectilinear stereo configuration with optical centres at $\mathbf{C}_L$ and $\mathbf{C}_R$. The scene contains two planar surfaces, with the light stripe projected onto the right-hand surface. The surface causes

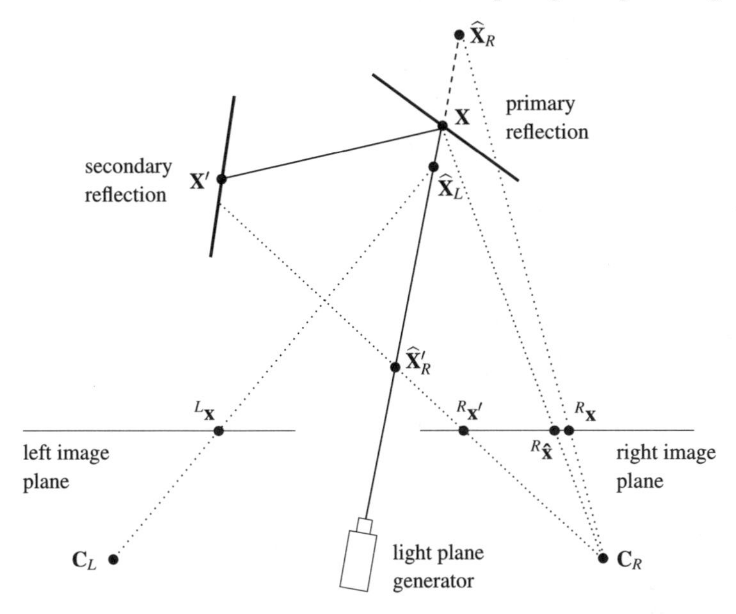

Fig. 3.2. Validation/reconstruction problem. *Reprinted from [151]. ©2004, Sage Publications. Used with permission.*

a primary reflection at $\mathbf{X}$ that is measured (using a noisy process) at $^L\mathbf{x}$ and $^R\mathbf{x}$ on the stereo image planes. However, a secondary specular reflection causes another stripe to appear at $\mathbf{X}'$, which is measured on the right image plane at $^R\mathbf{x}'$ but obscured from the left camera (in practice, such noisy measurements are produced by a variety of mechanisms other than secondary reflections). The 3D reconstructions, labelled $\widehat{\mathbf{X}}_L$, $\widehat{\mathbf{X}}_R$ and $\widehat{\mathbf{X}}'_R$ in Figure 3.2, are recovered as the intersection of the light plane and the rays back-projected through the image plane measurements. These points will be referred to as the *single-camera reconstructions*. As a result of noise on the CCD (exaggerated in this example), the back-projected rays do not intersect the physical reflections at $\mathbf{X}$ and $\mathbf{X}'$.

The robust scanning problem may now be stated as follows: given the laser plane position and the measurements $^L\mathbf{x}$, $^R\mathbf{x}$ and $^R\mathbf{x}'$, one of the left/right candidate pairs, ($^L\mathbf{x}$,$^R\mathbf{x}$) or ($^L\mathbf{x}$,$^R\mathbf{x}'$), must be chosen as representing stereo measurements of the primary reflection. Alternatively, all candidates may be rejected. This task is referred to as the *validation problem*, and a successful solution in this example should identify ($^L\mathbf{x}$,$^R\mathbf{x}$) as the valid measurements. The measurements should then be combined to estimate the position of the ideal projection $^R\hat{\mathbf{x}}$ (arbitrarily chosen to be on the right image plane) of the actual point $\mathbf{X}$ on the surface of the target.

Formulation of optimal validation/reconstruction algorithms should take account of measurement noise, which is not correctly modelled in previous related work. In [155] and [109], laser stripe measurements are validated by applying a *fixed* threshold to the difference between corresponding single-camera reconstructions ($\widehat{\mathbf{X}}_L$, $\widehat{\mathbf{X}}_R$

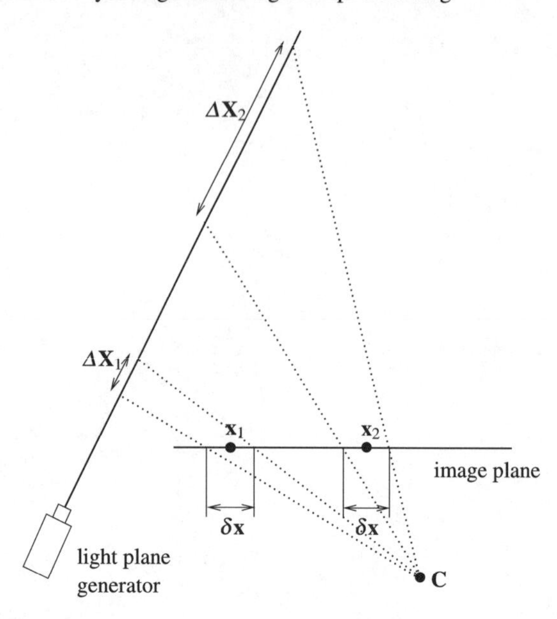

Fig. 3.3. Variation of reconstruction error with depth. *Reprinted from [151]. ©2004, Sage Publications. Used with permission.*

and $\widehat{\mathbf{X}}'_R$ in Figure 3.2). Such a comparison requires a uniform reconstruction error over all depths, which Figure 3.3 illustrates is clearly not the case. Two independent measurements at $\mathbf{x}_1$ and $\mathbf{x}_2$ generally exhibit a constant error variance on the image plane, as indicated by the interval $\delta\mathbf{x}$. However, projecting $\delta\mathbf{x}$ onto the laser plane reveals that the reconstruction error increases with depth, since $\Delta\mathbf{X}_1 < \Delta\mathbf{X}_2$ in Figure 3.3. Thus, the validation threshold on depth difference should increase with depth to account for measurement noise, otherwise validation is more lenient for closer reconstructions. Similarly, taking either $\widehat{\mathbf{X}}_L$, $\widehat{\mathbf{X}}_R$ or the arithmetic average $\frac{1}{2}(\widehat{\mathbf{X}}_L+\widehat{\mathbf{X}}_R)$ as the final reconstruction in Figure 3.2 is generally sub-optimal for noisy measurements.

The following sections present optimal solutions to the validation/reconstruction problem, based on an error model with the following features (the assumptions of the error model are corroborated with experimental results in Section 3.5.2):

1. Light stripe measurement errors are independent and Gaussian distributed with uniform variance over the entire image plane.
2. The dominant error in the light plane position is the angular error about the axis of rotation as the plane is scanned across the target, which couples all measurements in a given image.
3. All other parameters of the sensor (described in the following section), are assumed to be known with sufficient accuracy that any uncertainty can be ignored for the purpose of validation and reconstruction.

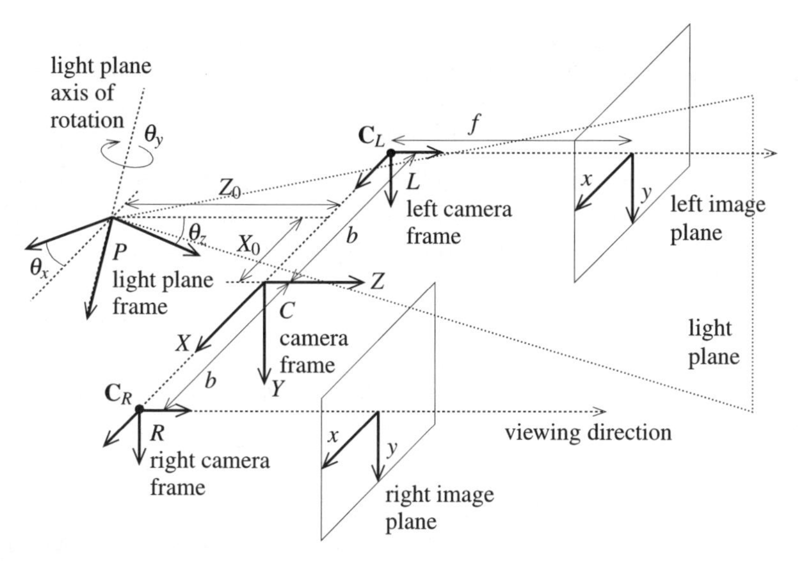

Fig. 3.4. Light stripe camera system model. *Reprinted from [151]. ©2004, Sage Publications. Used with permission.*

3.2.2 System Model

Figure 3.4 details the parameters of the system model for the stereoscopic light stripe sensor. L and R denote the left and right camera frames, P is the frame of the rotating light plane and C is the camera frame (defined in Section 2.2.3). As described in Section 2.2.1, the cameras are modelled by the 3×4 projection matrices ${}^{L,R}\mathrm{P}$, given by equation (2.29). The cameras are assumed to have identical focal length f, and are in rectilinear stereo configuration (optical axes aligned to the world z-axis) with optical centres located at $\mathbf{C}_{L,R} = (\mp b, 0, 0)^\top$ in the camera frame. As described in Section 2.2.4, the cameras are allowed to verge about the y-axis, and projective rectification is applied to every frame to recover the equivalent rectilinear stereo measurements.

Frame P is rigidly attached to the laser, and points $\mathbf{X}$ on the light plane are defined by the plane equation $\boldsymbol{\Omega}^\top \mathbf{X} = 0$, where $\boldsymbol{\Omega}$ represents the parameters of the light plane. Furthermore, frame P is defined such that the light plane is approximately vertical and parallel to the z-axis. This allows $\boldsymbol{\Omega}$ to be expressed in P as

$$ {}^{P}\boldsymbol{\Omega} = (1,\ B_0,\ 0,\ D_0)^\top \tag{3.1} $$

where B_0 is related to the small angle between the plane and the y-axis ($B_0 \ll 1$ for an approximately vertical plane), and D_0 is the distance of the plane from the origin. Thus, the normal vector of the plane is already normalized, since $|(1, B_0, 0)| \approx 1$. During a scan, frame P rotates about its y-axis with angle θ_y, where $\theta_y \equiv 0$ when the light plane is parallel to the optical axes of the cameras. The rotation axis intersects the xz-plane of the camera frame at $(X_0, 0, Z_0)^\top$ (in the camera frame), and the orientation of the rotation axis relative to the y-axis of the camera frame is defined by the small fixed angles θ_x and θ_z. The scan angle (θ_y in Figure 3.4) is assumed to have a linear relationship with the measured optical encoder value e via two additional parameters m and k:

$$\theta_y = me + k \tag{3.2}$$

Now let ${}^C\mathrm{H}_P$ represent the homogeneous coordinate transformation from P to C. Representing the homogeneous transformation for rotation about the x, y and z-axes by angle θ by matrices $\mathrm{R}_x(\theta)$, $\mathrm{R}_y(\theta)$ and $\mathrm{R}_z(\theta)$, and the translation of $(X,Y,Z)^\top$ by the matrix $\mathrm{T}(X,Y,Z)$, ${}^P\mathrm{H}_C$ can be expressed as

$${}^C\mathrm{H}_P = \mathrm{T}(X_0, 0, Z_0) \cdot \mathrm{R}_z(\theta_z) \cdot \mathrm{R}_x(\theta_x) \cdot \mathrm{R}_y(\theta_y) \tag{3.3}$$

It is straightforward to show that if ${}^P\mathrm{H}_C$ is the coordinate transformation from P to C, the plane parameters transform from P to C as

$${}^C\boldsymbol{\Omega} = ({}^C\mathrm{H}_P)^{-\top} \cdot {}^P\boldsymbol{\Omega} \tag{3.4}$$

Combining equations (3.3) and (3.4), the laser plane parameters are expressed in the camera frame as:

$${}^C\boldsymbol{\Omega} = \begin{pmatrix} c_y c_z - s_x s_y s_z - B_0 c_x s_z \\ c_y s_z + s_x s_y c_z + B_0 c_x c_z \\ -c_x s_y + B_0 s_x \\ (s_x s_y s_z - c_y c_z + B_0 c_x s_z) X_0 + (c_x s_y - B_0 s_x) Z_0 + D_0 \end{pmatrix} \tag{3.5}$$

where $c_x = \cos\theta_x$ and $s_x = \sin\theta_x$. By making the simplifying assumptions B_0, θ_x, $\theta_z \ll 1$, many insignificant terms in equation (3.5) can be neglected to give an approximate model:

$${}^C\boldsymbol{\Omega} = \begin{pmatrix} \cos\theta_y \\ \theta_x \sin\theta_y + \theta_z \cos\theta_y + B_0 \\ -\sin\theta_y \\ -X_0 \cos\theta_y + Z_0 \sin\theta_y + D_0 \end{pmatrix} \tag{3.6}$$

Finally, equations (3.2) and (3.6) allow points on the laser plane to be identified in the world frame, using the plane equation ${}^C\boldsymbol{\Omega}^\top(B_0, D_0, X_0, Z_0, \theta_x, \theta_y, \theta_z)\,{}^C\mathbf{X} = 0$.

3.2.3 Generalized Image Plane Error Function

This section now presents an analytical treatment of the validation/reconstruction problem. Let ${}^L\mathbf{x}$ and ${}^R\mathbf{x}$ represent noisy measurements of the stripe on corresponding epipolar lines, and let $\boldsymbol{\Omega} = (A, B, C, D)^\top$ represent the location of the light plane. The likelihood that the measurements correspond to the primary reflection of the stripe can be formulated as an error distance on the image plane. Let $\widehat{\mathbf{X}}$ represent the maximum likelihood 3D reconstruction, which is constrained to coincide with the laser plane, but not necessarily at the intersection of the back-projected rays from the noisy image plane measurements. Errors in the light plane parameters are considered in the next section, but for now the light plane parameters $\boldsymbol{\Omega}$ are assumed to be known exactly. The optimal reconstruction $\widehat{\mathbf{X}}$ projects to the ideal corresponding points on the left and right image planes at ${}^{L,R}\hat{\mathbf{x}} = {}^{L,R}\mathrm{P}\widehat{\mathbf{X}}$, according to equation (2.23). Now, in a similar manner to equation (6.12), the sum of squared errors between the ideal and measured points can be used to determine whether the candidate measurement pair $(\mathbf{x}_L, \mathbf{x}_R)$ corresponds to a point $\widehat{\mathbf{X}}$ on the light plane:

$$\begin{aligned} E &= d^2({}^L\mathbf{x}, {}^L\hat{\mathbf{x}}) + d^2({}^R\mathbf{x}, {}^R\hat{\mathbf{x}}) \\ &= d^2({}^L\mathbf{x}, {}^L\mathrm{P}\widehat{\mathbf{X}}) + d^2({}^R\mathbf{x}, {}^R\mathrm{P}\widehat{\mathbf{X}}) \end{aligned} \tag{3.7}$$

where $d(\mathbf{x}_1, \mathbf{x}_2)$ is the Euclidean distance between $\mathbf{x}_1$ and $\mathbf{x}_2$. For a given candidate pair, the optimal reconstruction $\widehat{\mathbf{X}}$ with respect to image plane error is found by a constrained minimization of E with respect to the condition that $\widehat{\mathbf{X}}$ is on the laser plane:

$$\boldsymbol{\Omega}^\top \widehat{\mathbf{X}} = 0 \tag{3.8}$$

When multiple ambiguous correspondences exist, equation (3.7) is optimized with respect to the constraint in (3.8) for all possible candidate pairs, and the pair with minimum error is chosen as the most likely correspondence. Finally, the result is validated by imposing a threshold on the maximum allowed squared image plane error E.

Performing the constrained optimization of equations (3.7)-(3.8) is analytically cumbersome. Fortunately, the problem may be reduced to an unconstrained optimization by determining the direct relationship between projections ${}^L\hat{\mathbf{x}}$ and ${}^R\hat{\mathbf{x}}$ for points on the light plane. Taking the intersection between the light plane and the back-projected ray from ${}^R\hat{\mathbf{x}}$, the relationship between $\widehat{\mathbf{X}}$ and ${}^R\hat{\mathbf{x}}$ for points on the light plane $\boldsymbol{\Omega}$ is given by (see Appendix B for a complete derivation of this result):

$$\widehat{\mathbf{X}} = [\mathbf{C}_R ({}^R\mathrm{P}^{+\,R}\hat{\mathbf{x}})^\top - ({}^R\mathrm{P}^{+\,R}\hat{\mathbf{x}})\mathbf{C}_R^\top]\boldsymbol{\Omega} \tag{3.9}$$

where ${}^R\mathrm{P}^+$ is the pseudo-inverse of ${}^R\mathrm{P}$ (given by equation (2.35)). Now, projecting $\widehat{\mathbf{X}}$ onto the left image plane, the relationship between the projections ${}^L\hat{\mathbf{x}}$ and ${}^R\hat{\mathbf{x}}$ for points on the light plane $\boldsymbol{\Omega}$ can be expressed as (see Appendix B):

$$\begin{aligned} {}^L\hat{\mathbf{x}} &= {}^L\mathrm{P}{}^R\hat{\mathbf{x}} \\ &= \left({}^L\mathrm{P}[\mathbf{C}_R\boldsymbol{\Omega}^\top - (\mathbf{C}_R^\top\boldsymbol{\Omega})\mathrm{I}]{}^R\mathrm{P}^+\right){}^R\hat{\mathbf{x}} \end{aligned} \tag{3.10}$$

Equation (3.10) is of the form ${}^L\hat{\mathbf{x}} = \mathrm{H}{}^R\hat{\mathbf{x}}$ and simply states that points on the laser plane induce a homography between coordinates on the left and right image planes, which is consistent with known results [52]. Finally, the error function becomes

$$E = d^2({}^L\mathbf{x}, \mathrm{H}{}^R\hat{\mathbf{x}}) + d^2({}^R\mathbf{x}, {}^R\hat{\mathbf{x}}) \tag{3.11}$$

where $\mathrm{H} = {}^L\mathrm{P}[\mathbf{C}_R\boldsymbol{\Omega}^\top - (\mathbf{C}_R^\top\boldsymbol{\Omega})\mathrm{I}]{}^R\mathrm{P}^+$. The reconstruction problem can now be formulated as an unconstrained optimization of equation (3.11) with respect to ${}^R\hat{\mathbf{x}}$. Then, the minimum squared error E over all candidates is used to resolve the validation/correspondence problem, and the reconstruction $\widehat{\mathbf{X}}$ can be recovered from (3.9).

3.2.4 Special Case: Rectilinear Stereo and Pin-Hole Cameras

The results of the previous section apply to general camera models and stereo geometry. However, the special case of rectilinear stereo and pin-hole cameras is important as it reduces equation (3.11) to a single degree of freedom. Furthermore, rectilinear stereo applies without loss of generality (after projective rectification), and the pin-hole model is a good approximation for CCD cameras (after correcting for radial lens distortion). The stereo cameras used in this work are assumed to have unit aspect ratio and no skew (see Section 2.2.1), and the pin-hole models are parameterized by equal focal length f. Details of the analysis presented in this section can be found in Appendix B.

The camera centres $\mathbf{C}_{L,R}$ and projection matrices ${}^{L,R}\mathrm{P}$ for rectilinear pin-hole cameras are given by equations (2.28)-(2.29). Substituting these into equation (3.10), the relationship between the projections of a point on the light plane can be written as:

$$ {}^{L}\hat{\mathbf{x}} = \begin{pmatrix} Ab-D & 2Bb & 2Cbf \\ 0 & -(Ab+D) & 0 \\ 0 & 0 & -(Ab+D) \end{pmatrix} {}^{R}\hat{\mathbf{x}} \tag{3.12} $$

In inhomogeneous coordinates, the relationship between ${}^{L}\hat{\mathbf{x}} = ({}^{L}\hat{x}, {}^{L}\hat{y})^{\top}$ and ${}^{R}\hat{\mathbf{x}} = ({}^{R}\hat{x}, {}^{R}\hat{y})^{\top}$ given by the homogeneous transformation in equation (3.12) can be expressed as

$$ {}^{L}\hat{x} = -\frac{(Ab-D){}^{R}\hat{x} + 2Bb{}^{R}\hat{y} + 2Cbf}{Ab+D} \tag{3.13} $$

$$ {}^{L}\hat{y} = {}^{R}\hat{y} \tag{3.14} $$

Since the axes of L and R are parallel (rectilinear stereo), the notation $\hat{y} \equiv {}^{L}\hat{y} = {}^{R}\hat{y}$ replaces equation (3.14). Rectilinear stereo gives rise to epipolar lines that are parallel to the x-axis, so the validation algorithm need only consider possible correspondences on matching scan-lines in the stereo images. Any measurement error in the stripe detection process (see Section 3.4.1) is assumed to be in the x-direction only, while the y-coordinate is fixed by the height of the scan-line. Thus, the y-coordinate of the optimal projections are also fixed by the scan-line, ie. $\hat{y} = y$, where y is the y-coordinate of the candidate measurements ${}^{L}\mathbf{x}$ and ${}^{R}\mathbf{x}$.

Finally, substituting equations (3.13)-(3.14) with $\hat{y} = y$ into (3.11), the image plane error E can be expressed as a function of a *single* variable, ${}^{R}\hat{x}$:

$$ E = ({}^{L}x + \alpha{}^{R}\hat{x} + \beta y + \gamma f)^2 + ({}^{R}x - {}^{R}\hat{x})^2 \tag{3.15} $$

where the following change of variables is introduced:

$$ \alpha = (Ab-D)/(Ab+D) \tag{3.16} $$

$$ \beta = 2Bb/(Ab+D) \tag{3.17} $$

$$ \gamma = 2Cb/(Ab+D) \tag{3.18} $$

For the experimental scanner, with $\boldsymbol{\Omega}$ given by equation (3.6), α, β and γ can be written as:

$$\alpha = -\frac{k_1 \cos\theta_y + k_2 \sin\theta_y + k_3}{\cos\theta_y + k_2 \sin\theta_y + k_3} \tag{3.19}$$

$$\beta = \frac{(1-k_1)(\theta_x \sin\theta_y + \theta_z \cos\theta_y + B_0)}{\cos\theta_y + k_2 \sin\theta_y + k_3} \tag{3.20}$$

$$\gamma = \frac{(k_1 - 1)\sin\theta_y}{\cos\theta_y + k_2 \sin\theta_y + k_3} \tag{3.21}$$

where $\theta_y = me + c$ and the following change of variables is made in the system parameters:

$$k_1 = -(b + X_0)/(b - X_0) \tag{3.22}$$

$$k_2 = Z_0/(b - X_0) \tag{3.23}$$

$$k_3 = D_0/(b - X_0) \tag{3.24}$$

Optimization of equation (3.15) now proceeds using standard techniques, setting $\frac{dE}{d^R\hat{x}} = 0$ and solving for ${}^R\hat{x}$. Let ${}^R\hat{x}^*$ represent the optimal projection resulting in the minimum squared error, E^*. It is straightforward to show (see Appendix B) that the optimal projection is given by

$${}^R\hat{x}^* = [{}^Rx - \alpha({}^Lx + \beta y + \gamma f)]/(\alpha^2 + 1) \tag{3.25}$$

and the minimum squared error E^* for the optimal solution is:

$$E^* = ({}^Lx + \alpha {}^Rx + \beta y + \gamma f)^2/(\alpha^2 + 1) \tag{3.26}$$

For completion, substituting equation (3.25) and ${}^R\hat{y}^* = y$ into (3.13) gives the corresponding optimal projection on the left image plane as

$${}^L\hat{x}^* = [\alpha^{2L}x - \alpha^R x - (\beta y + \gamma f)]/(\alpha^2 + 1) \tag{3.27}$$

Finally, the optimal 3D reconstruction $\widehat{\mathbf{X}}^*$ is recovered by substituting (3.25) into equation (3.9). In non-homogeneous coordinates, the optimal reconstruction leading to the minimum image plane error for candidate measurements ${}^L\mathbf{x}$ and ${}^R\mathbf{x}$ is (see Appendix B):

$$\widehat{X}^* = \frac{[(\alpha - 1)(\alpha^L x - {}^R x) - (\alpha + 1)(\beta y + \gamma f)]b}{(\alpha + 1)(\alpha^L x - {}^R x) + (\alpha - 1)(\beta y + \gamma f)} \tag{3.28}$$

$$\widehat{Y}^* = \frac{2by(\alpha^2 + 1)}{(\alpha + 1)(\alpha^L x - {}^R x) + (\alpha - 1)(\beta y + \gamma f)} \tag{3.29}$$

$$\widehat{Z}^* = \frac{2bf(\alpha^2 + 1)}{(\alpha + 1)(\alpha^L x - {}^R x) + (\alpha - 1)(\beta y + \gamma f)} \tag{3.30}$$

The validation problem can now be solved by evaluating E^* in equation (3.26) for all pairs of candidate measurements on matching scan-lines, and selecting the pair with the minimum error (less than some validation threshold). Once the valid measurements have been identified, the position of the light plane is calculated from the encoder count e using equations (3.2) and (3.19)-(3.21), and the optimal reconstruction is recovered from the image plane measurements using (3.28)-(3.30).

3.2.5 Laser Plane Error

The above solution is optimal with respect to the error of image plane measurements, and assumes that the parameters of the laser plane are known exactly. In practice, the encoder measurements are likely to suffer from both random and systematic error due to acquisition delay and quantization. Unlike the image plane error, the encoder error is constant for all stripe measurements in a given frame and thus cannot be minimized independently for candidate measurements on each scan-line.

Let ${}^{L}\mathbf{x}_i$ and ${}^{R}\mathbf{x}_i$, $i = 1 \ldots n$ represent valid corresponding measurements of the laser stripe on the n scan-lines in a frame. The reconstruction error $E_i^*(e)$ for each pair can be treated as a function of the encoder count via the system model in equations (3.19)-(3.21). The total error $E_{\text{tot}}^*(e)$ over all scan-lines for a given encoder count e is calculated as:

$$E_{\text{tot}}^*(e) = \sum_{i=1}^{n} E_i^*(e) \tag{3.31}$$

Finally, an optimal estimate of the encoder count e^* is calculated from the minimization

$$e^* = \arg\min_{e}[E_{\text{tot}}^*(e)] \tag{3.32}$$

Since $E_{\text{tot}}^*(e)$ is a non-linear function, the optimization in equation (3.32) is implemented numerically using Levenberg-Marquardt (LM) minimization from MINPACK [105], with the measured value of e as the initial estimate.

As noted above, valid corresponding measurements must be identified before calculating $E_{\text{tot}}^*(e)$. However, since the correspondences are determined by minimizing E^* over all candidate pairs given the plane parameters, the correspondences are also a function of the encoder count. Thus, the refined estimate e^* may relate to a different set of optimal correspondences than those from which it was calculated. To resolve this issue, the optimal correspondences and encoder count are calculated recursively. In the first iteration, correspondences are calculated using the measured encoder value e to yield the initial estimate e_0^* via equations (3.31)-(3.32). A new set of correspondences are then extracted from the raw measurements using the refined encoder value e_0^*. If the new correspondences differ from the previous iteration, an updated estimate of the encoder value e_1^* is calculated (using e_0^* as the initial guess). The process is repeated until a stable set of correspondences is found.

The above process is applied to each captured frame, and the optimal encoder count e^* and valid correspondences, ${}^{L}\mathbf{x}_i$ and ${}^{R}\mathbf{x}_i$, are substituted into equations (3.28)-(3.30) to finally recover the optimal 3D profile of the laser.

3.2.6 Additional Constraints

As already described, robust stripe detection is based on minimization of the image plane error in equation (3.26). However, the minimum image plane error is a necessary but insufficient condition for identifying valid stereo measurements. In the practical implementation, two additional constraints are employed to improve the robustness of stripe detection.

The first constraint simply requires stripe candidates to be moving features; a valid measurement must not appear at the same position in previous frames. This is implemented by processing only those pixels with sufficiently large intensity difference between successive frames. While this constraint successfully rejects static stripe-like edges or textures in most scenes, it has little effect on cross-talk or reflections, since these also appear as moving features.

The second constraint is based on the fact that valid measurements only occur within a sub-region of the image plane, depending on the angle of the light plane. It can be shown (see equation (B.29) in Appendix B), that the inhomogeneous z-coordinate of a single-camera reconstruction $\widehat{\mathbf{X}}$ can be expressed as a function of the image plane projections ${}^{L,R}\hat{\mathbf{x}}$ as

$$\widehat{Z} = \pm \frac{2bf}{(\alpha+1)\,{}^{L,R}\hat{x} + \beta y + \gamma f} \tag{3.33}$$

where the positive sign is taken for L and negative sign for R. Rearranging the above, the projected x-coordinate of a point on the light plane may be expressed as a function of depth $\widehat{Z}$ and the height y of the scan-line:

$${}^{L,R}\hat{x} = -\frac{\beta y + \gamma f}{\alpha+1} \pm \frac{2bf}{\widehat{Z}(\alpha+1)} \tag{3.34}$$

The extreme boundaries for valid measurements can now be found by taking the limit of equation (3.34) for points on the light plane near and far from the camera. Taking the limit for distant reflections gives one boundary at:

$$\lim_{\widehat{Z}\to\infty} {}^{L,R}\hat{x} = -\frac{\beta y + \gamma f}{\alpha+1} \tag{3.35}$$

Taking the limit $\widehat{Z} \to 0$ for close reflections gives the other boundary at ${}^{L,R}\hat{x} \to \pm\infty$. Now, if w is the width of the captured image, valid measurements on a scan-line at height y must be constrained to the x-coordinate ranges

$${}^{L}x \in \left[-\frac{\beta v + \gamma f}{\alpha+1}, +\frac{w}{2}\right] \tag{3.36}$$

$${}^{R}x \in \left[-\frac{w}{2}, -\frac{\beta v + \gamma f}{\alpha+1}\right] \tag{3.37}$$

Stripe extraction is only applied to pixels within the boundaries defined in (3.36)-(3.37); pixels outside these ranges are immediately classified as invalid. In addition to improving robustness, sub-region processing also reduces computational expense by halving the quantity of raw image data and decreasing the number of stripe candidates tested for correspondence.

3.3 Active Calibration of System Parameters

In this section, determination of the unknown parameters in the model of the light stripe scanner are considered. Let the unknown parameters be represented by the vector

$$\mathbf{p} = (k_1, k_2, k_3, \theta_x, \theta_z, B_0, m, k)$$

where k_1, k_2 and k_3 were introduced in equations (3.22)-(3.24). Since most of the parameters relate to mechanical properties, the straightforward approach to calibration is manual measurement. However, such an approach would be both difficult and increasingly inaccurate as parameters vary through mechanical wear. To overcome this problem, a strategy is now proposed to optimally estimate **p** using only image-based measurements of a non-planar but otherwise arbitrary surface with favourable reflectance properties (the requirement of non-planarity is discussed below). This allows calibration to be performed cheaply and during normal operation.

The calibration procedure begins by scanning the stripe across the target and recording the encoder and image plane measurements for each captured frame. Since the system parameters are initially unknown, the validation problem is approximated by recording only the brightest pair of features per scan-line. Let ${}^L\mathbf{x}_{ij}$ and ${}^R\mathbf{x}_{ij}$, $i = 1 \ldots n_j$, $j = 1 \ldots t$ represent the centroids of the brightest corresponding features on n_j scan-lines of t captured frames, and let e_j represent the measured encoder value for each frame. As described earlier, image plane measurements have independent errors, while the encoder error couples all measurements in a given frame. Thus, optimal system parameters are determined from iterative minimization of the stripe measurement and encoder errors, based on the algorithm first described in Section 3.2.5. First, the total image plane error E^*_{tot} is summed over all frames:

$$E^*_{\text{tot}} = \sum_{j=1}^{t} \sum_{i=1}^{n_j} E^*({}^L\mathbf{x}_{ij}, {}^R\mathbf{x}_{ij}, e_j, \mathbf{p}) \tag{3.38}$$

where E^* is defined in equation (3.26). The requirement of a non-planar calibration target can now be justified. For a planar target, the stripe appears as a straight line and the image plane measurements obey a linear relationship of the form $x_{ij} = a_j y_{ij} + b_j$. Then, the total error E^*_{tot} reduces to the form

$$E^*_{\text{tot}} = \sum_{j=1}^{t} \sum_{i=1}^{n_j} (A_j y_{ij} + B_j)^2 \tag{3.39}$$

Clearly, the sign of A_j and B_j cannot be determined from equation (3.39), since the total error remains unchanged after substituting $-A_j$ and $-B_j$. The geometrical interpretation of this result is illustrated in Figure 3.5, which shows the 2D analogue of a planar target scan. For any set of encoder values e_j and collinear points $\mathbf{X}_j$ measured over t captured frames, there exist two symmetrically opposed laser plane generators capable of producing identical results. This ambiguity can be overcome by constraining the calibration target to be non-planar. It may also be possible for certain non-planar targets to produce ambiguous results, but the current implementation assumes that such an object will rarely be encountered.

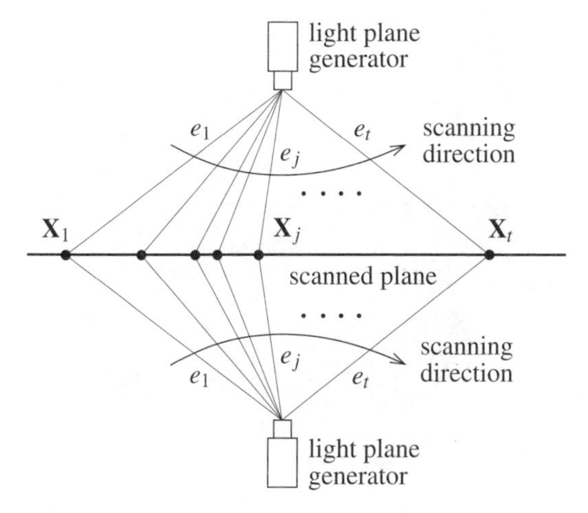

Fig. 3.5. Possible positions of the light plane from the scan of a planar calibration target. The position of the light plane generator is ambiguous. *Reprinted from [151]. ©2004, Sage Publications. Used with permission.*

An initial estimate $\mathbf{p}_0^*$ of the parameter vector is given by the minimization

$$\mathbf{p}_0^* = \arg\min_{\mathbf{p}}[E_{\text{tot}}^*(\mathbf{p})] \tag{3.40}$$

using the measured encoder values e_j and stereo correspondences ${}^L\mathbf{x}_{ij}$ and ${}^R\mathbf{x}_{ij}$. Again, equation (3.40) is implemented numerically using LM minimization. The stripe measurements ${}^L\mathbf{x}_{ij}$ and ${}^R\mathbf{x}_{ij}$ are likely to contain gross errors resulting from the initial coarse validation constraint (in the absence of known system parameters). Thus, the next calibration step refines the measurements by applying outlier rejection. Using e_j and the initial estimate $\mathbf{p}_0^*$, the image plane error $E^*({}^L\mathbf{x}_{ij}, {}^R\mathbf{x}_{ij}, e_j, \mathbf{p}_0^*)$ in equation (3.26) is calculated for each stereo pair. The measurements are then sorted in order of increasing error, and the top 20% are discarded.

The system parameters and encoder values are then sequentially refined in an iterative process. The initial estimate $\mathbf{p}_0^*$ is only optimal with respect to image plane error, assuming exact encoder values e_j. To account for encoder error, the encoder value is refined for each frame using the method described in Section 3.2.5 with the initial estimate $\mathbf{p}_0^*$ of the system model. The resulting encoder estimates $e_{j,0}^*$ are optimal with respect to $\mathbf{p}_0^*$. A refined system model $\mathbf{p}_1^*$ is then obtained from equation (3.40) using the latest encoder estimates $e_{j,0}^*$ and inlier image plane measurements. At the kth iteration, the model is considered to have converged when the fractional change in total error E_{tot}^* is less then a threshold δ:

$$\frac{E_{\text{tot},k-1}^* - E_{\text{tot},k}^*}{E_{\text{tot},k-1}^*} < \delta \tag{3.41}$$

The final parameter vector $\mathbf{p}_k^*$ is stored as the near-optimal system model for processing regular scans using the methods described in Section 3.2. A final check for global

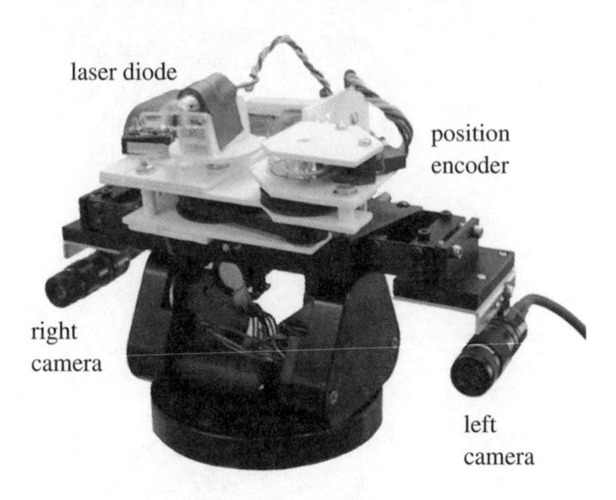

Fig. 3.6. Experimental stereoscopic light stripe scanner. *Reprinted from [151]. ©2004, Sage Publications. Used with permission.*

optimality is performed by comparing the minimum total error $E^*_{\text{tot},k}$ to a fixed threshold, based on an estimate of the image plane error. The rare case of non-convergence (less than 10% of trials) is typically due to excessive outliers introduced by the sub-optimal maximum intensity validation constraint applied to the initial measurements. Non-convergence is resolved by repeating the calibration process with a new set of data.

The calibration technique presented here is practical, fast and accurate, requiring only a single scan of any suitable non-planar scene. Furthermore, the method does not rely on measurement or estimation of any metric quantities, and so does not require accurate knowledge of camera parameters b and f. Thus, image-based calibration allows the validation and correspondence problems to be solved robustly and independently of reconstruction accuracy.

3.4 Implementation

Figure 3.6 shows the components of the experimental stereoscopic light stripe scanner, which is mounted on the Biclops head. A vertical light plane is generated by a laser diode module with a cylindrical lens, and is scanned across a scene by rotating the laser about a vertical axis. The angle of rotation is measured by an optical encoder connected to the output shaft via a toothed belt. PAL colour cameras capture stereo images of the stripe at 384×288 pixel (half-PAL) resolution. The shaft encoder and stereo images are recorded at regular 40 ms intervals (25 Hz PAL frame-rate). The laser is mechanically geared to displace the stripe by about one pixel of horizontal motion per captured frame, so a complete scan requires approximately 384 processed frames (15 seconds).

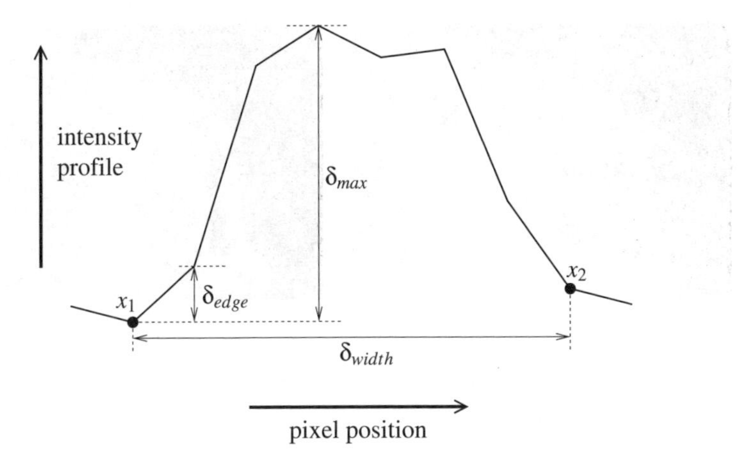

Fig. 3.7. Thresholds for robust extraction of multi-modal pulses. *Reprinted from [151].*

For each captured stereo pair, the vertical stripe is extracted from the image plane and combined with the laser angle and system parameters to recover the 3D profile of the illuminated surface. As the stripe is scanned across the scene, the laser profiles are assembled into an array of 3D points, which is referred to as the *range map*. Each element of the range map records the 3D position of the target surface as viewed from the right camera. A colour image is captured and registered with the range map at the completion of a scan. Captured frames are processed at PAL frame-rate (25 Hz) on the 2.2 GHz dual Xeon host PC. Motor control and optical encoder measurements are implemented on a PIC microcontroller, which communicates with the host PC via an RS-232 serial link (see Figure 7.2).

3.4.1 Light Stripe Measurement

Laser stripe extraction is performed using intensity data only, which is calculated by taking the average of the colour channels. As noted in Section 3.2.6, the motion of the stripe distinguishes it from the static background, which is eliminated by subtracting the intensity values in consecutive frames and applying a minimum difference threshold. The resulting *difference image* is morphologically eroded and dilated to reduce noise and improve the likelihood of stripe detection. In Section 3.2.6 it was also shown that valid measurements occur in a predictable sub-region of the image. This is calculated from equations (3.36)-(3.37) and the measured encoder value, and pixels outside this region are set to zero in the difference image. Further processing is only applied to pixels with non-zero difference.

The intensity profile on each scan-line is then examined to locate candidate stripe measurements. If the stripe appeared as a simple unimodal pulse, the local maxima would be sufficient to detect candidates. However, mechanisms including sensor noise, surface texture and saturation of the CCD interfere and perturb the intensity

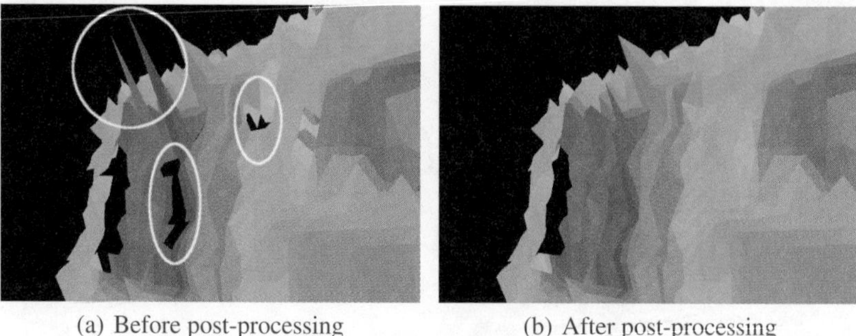

(a) Before post-processing (b) After post-processing

Fig. 3.8. Removal of impulse noise and holes (eliminated features are circled on the left). *Reprinted from [151]. ©2004, Sage Publications. Used with permission.*

profile. These issues are overcome by extracting pulses using a more sophisticated strategy of intensity edge extraction and matching. On each scan-line, left and right edges are identified as an increase or decrease in the intensity profile according to the thresholds defined in Figure 3.7. Processing pixels from left to right, the location of a left edge x_l is detected when the intensity difference between successive pixels exceeds a threshold δ_{edge}, and the closest local intensity maxima to the right of x_l exceeds the intensity at x_l by a larger threshold δ_{max}. Right edges x_r are extracted by processing the scan-line in reverse. Finally, the edges are examined to identify left/right pairs without intervening edges. When x_l and x_r are closer than a threshold distance δ_{width}, the pair are assumed to delimit a candidate pulse. The pulse centroid is calculated to sub-pixel accuracy as the mean pixel position weighted by the intensity profile within these bounds.

The result of the above process is a set of candidate stripe locations on each scan-line of the stereo images. Along with the measured encoder value, these candidates are analyzed using the techniques described in Section 3.2 to refine the laser plane angle, identify valid corresponding measurements and reconstruct an optimal 3D profile. The reconstruction on each scan-line is stored in the range map at the location of the corresponding measurement in the right image.

3.4.2 Range Data Post-processing

Post-processing is applied *after* each complete scan to further refine the measured data. Despite robust scanning, the raw range map may still contain outliers as the stripe validation conditions are occasionally satisfied by spurious noise. Fortunately, the sparseness of the outliers make them easy to detect and remove using a simple thresholding operation: the minimum distance between each 3D point and its eight neighbours is calculated, and when this exceeds a threshold (10 mm in the current implementation), the associated point is removed from the range map.

Holes (pixels for which range data could not be recovered) may occur in the range map due to specular reflections, poor surface reflectivity, random noise and outlier removal. A further post-processing step fills these gaps with interpolated depth data. Each empty pixel is checked to determine whether it is bracketed by valid data within a vertical or horizontal distance of two pixels. To avoid interpolating across depth discontinuities, the distance between the bracketing points must be less than a fixed threshold (30 mm in the current implementation). Empty pixels satisfying these constraints are assigned a depth linearly interpolated between the valid bracketing points. The effect of both outlier rejection and interpolation on a raw scan is demonstrated in Figure 3.8.

Finally, a colour image is registered with the range map. Since robust scanning allows the sensor to operate in normal light, the cameras used for stripe detection also capture colour information. However, depth and colour cannot be sampled simultaneously for any given pixel, since the laser masks the colour of the surface. Instead, a complete range map is captured before registering a colour image from the right camera (assuming the cameras have not moved). Each pixel in this final image yields the colour of the point measured in the corresponding pixel of the range map.

3.5 Experimental Results

3.5.1 Robustness

To evaluate the performance of the robust methods proposed in this chapter, the scanner is implemented along with two other common techniques on the same experimental platform. The first method is a simple single-camera scanner without any optimal or robust properties. A single-camera reconstruction is calculated from equation (3.9), using image plane measurements from the right camera only. Since no validation is possible in this configuration, the stripe is simply detected as the brightest feature on each scan-line. The second alternative implementation will be referred to as a *double-camera* scanner. This approach is based on the robust techniques proposed in [109, 155], which exploit the requirement of consensus between two independent single-camera reconstructions. Again, the single-camera reconstructions $\widehat{\mathbf{X}}_L$ and $\widehat{\mathbf{X}}_R$ are calculated from equation (3.9), from measurements ${}^R\mathbf{x}$ and ${}^R\mathbf{x}$. Measurements are discarded when $|\widehat{\mathbf{X}}_L - \widehat{\mathbf{X}}_R|$ exceeds a fixed distance threshold. For valid measurements, the final reconstruction is calculated as $\frac{1}{2}(\widehat{\mathbf{X}}_L + \widehat{\mathbf{X}}_R)$.

The performance of the three methods in the presence of a phantom stripe (secondary reflection) was assessed using the test scene shown in Figure 3.9. A mirror at the rear of the scene creates a reflection of the objects and scanning laser, simulating the effect of cross-talk and specular reflections. To facilitate a fair comparison, the three methods operated simultaneously on the same raw measurements captured during a single scan of the scene.

Figure 3.10 shows the colour/range data captured by the single-camera scanner. As a result of erroneous associations between the phantom stripe and laser plane, numerous phantom surfaces appear in the scan without any physical counterpart. Figure 3.11 shows the output of the double-camera scanner, which successfully removes the spurious surfaces. However, portions of real surfaces have also been rejected, since

Fig. 3.9. Robust scanning experiment. A mirror behind the objects simulates the effect of cross-talk and reflections. *Reprinted from [151]. ©2004, Sage Publications. Used with permission.*

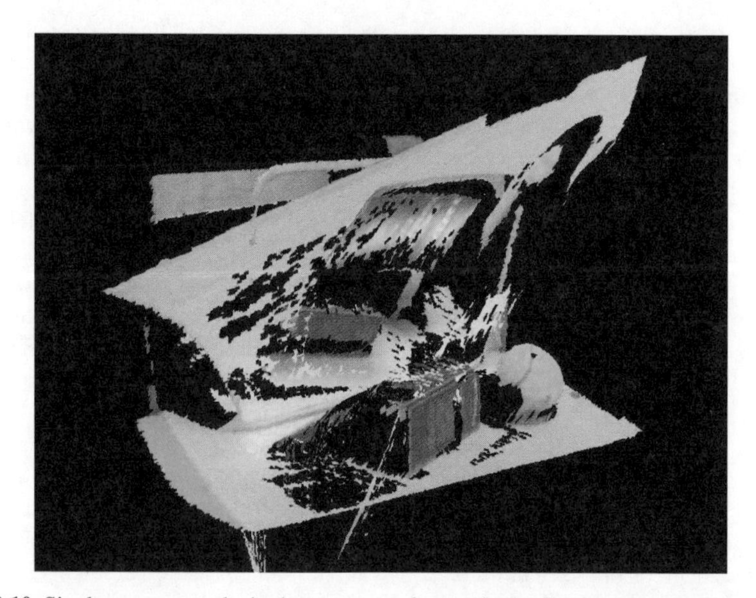

Fig. 3.10. Single-camera results in the presence of secondary reflections (see *Multimedia Extensions* for VRML model). *Reprinted from [151]. ©2004, Sage Publications. Used with permission.*

Fig. 3.11. Double-camera results in the presence of secondary reflections (see *Multimedia Extensions* for VRML model). *Reprinted from [151]. ©2004, Sage Publications. Used with permission.*

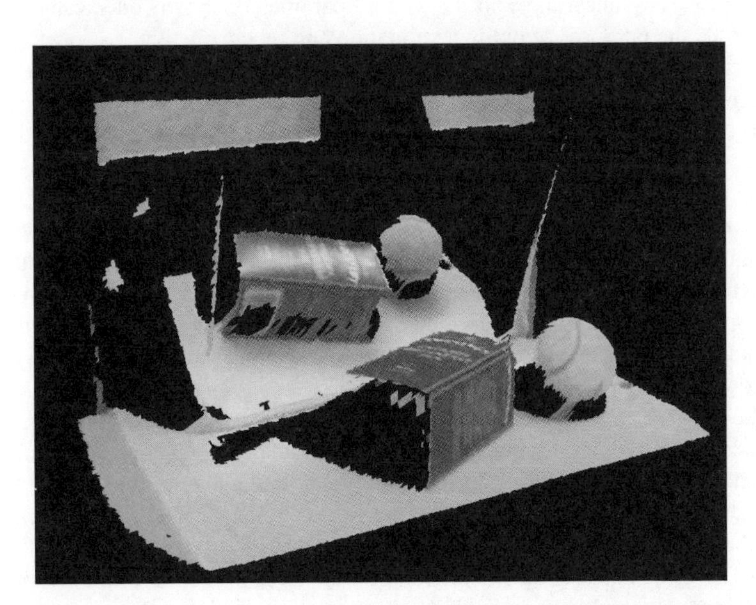

Fig. 3.12. Robust scanner results in the presence of secondary reflections (see *Multimedia Extensions* for VRML model). *Reprinted from [151]. ©2004, Sage Publications. Used with permission.*

the algorithm is unable to disambiguate the phantom stripe from the primary reflection when both appear in the scene. Finally, Figure 3.12 shows the result using the techniques presented in this chapter. The portions of the scene missing from Figure 3.11 are successfully detected using the proposed robust scanner, while the phantom stripe has been completely rejected. Also noteworthy is the implicitly accurate registration of colour and range. VRML models of the scans in Figures 3.10-3.12 are provided in the *Multimedia Extensions*.

The single-camera result highlights the need for robust methods when using light stripe scanners on a domestic robot. While the double-camera scanner successfully rejects reflections and cross-talk, the high rejection rate for genuine measurements may cause problems for segmentation or other subsequent processing. In contrast, segmentation and object classification have been successfully applied to the colour/range data from the proposed robust scanner to facilitate high-level domestic tasks (see Chapter 4 and [147]).

3.5.2 Error Analysis

The results in this section experimentally validate of the system and noise models used to derive the theoretical results. In particular, the encoder angle estimation and calibration techniques described in Sections 3.2.5 and 3.3 are shown to be sufficiently accurate that any uncertainty in the system parameters and encoder values can be ignored for the purposes of optimal validation and reconstruction.

First, the calibration procedure described in Section 3.3 was performed using the corner formed by two boxes as the calibration target, and the valid image plane measurements and encoder values for each frame were recorded. Using the estimated system parameters, the optimal projections ${}^{L,R}\hat{x}^*$ and residuals $({}^{L,R}x - {}^{L,R}\hat{x}^*)$ were calculated from equations (3.25) and (3.27) for all measurements. Figure 3.13 shows the histogram of residuals for measurements on the right image plane, and a Gaussian distribution with the same parameters for comparison. The residuals are approximately Gaussian distributed as expected, and assuming the light stripe measurement errors are similarly distributed, the error model proposed in Section 3.2.1 is found to be valid. The variance of the image plane measurements are shown in the first two rows of Table 3.1.

The variance of the system parameters and encoder values were determined using statistical bootstrapping. In this process, the residuals were randomly and uniformly sampled (with replacement) from the initial data, and added to the optimal projections ${}^{L,R}\hat{x}^*$ to generate a new set of pseudo-measurements. The system parameters were then estimated for each new set of pseudo-measurements using the calibration process described in Section 3.3. A total of 5000 re-sampling experiments were performed, and the resulting variance in the estimated system parameters are shown in the third column of Table 3.1.

Finally, the optimality of the proposed reconstruction method is assessed by calculating the contribution from each parameter to the variance of the optimal image plane reconstruction ${}^{R}\hat{x}^*$ in equation (3.25). Representing the components of the parameter vector as $\mathbf{p} = \{p_i\}$, and assuming the parameters have independent noise,

Table 3.1. Average and variance of measurements and system parameters, and contribution to reconstruction error.

p_i	$\bar{p}_i$	$\text{var}(p_i)$	$\text{var}({}^R\hat{x}^*)_i$
${}^Lx$	59.2	8×10^{-3}	2×10^{-3}
${}^Rx$	-47.4	7×10^{-3}	2×10^{-3}
y	6.0	0.0	0.0
e	451.5	5×10^{-5}	2×10^{-5}
k_1	-1.1033	2×10^{-8}	1×10^{-5}
k_2	0.1696	4×10^{-8}	2×10^{-8}
k_3	0.0717	1×10^{-8}	2×10^{-5}
θ_x	-0.0287	3×10^{-10}	2×10^{-12}
θ_z	0.0094	5×10^{-8}	2×10^{-6}
B_0	-0.0004	5×10^{-8}	2×10^{-6}
m	0.001105	7×10^{-16}	3×10^{-5}
c	-0.4849	1×10^{-10}	4×10^{-5}

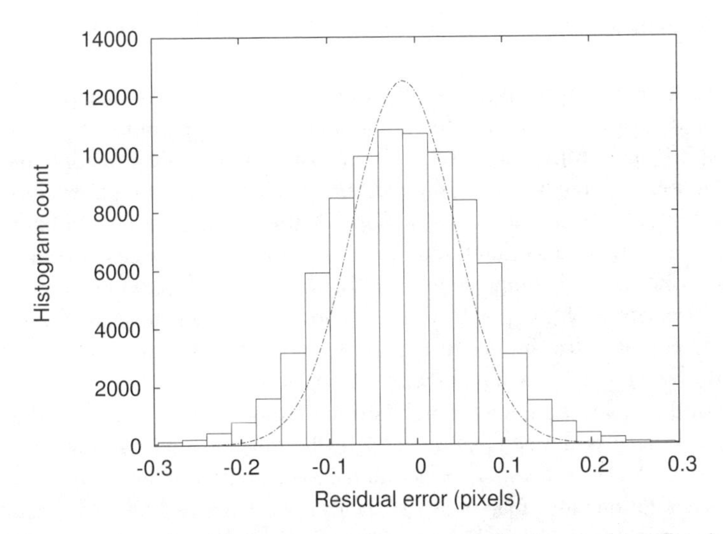

Fig. 3.13. Distribution of residual reconstruction errors on the image plane. *Reprinted from [151]. ©2004, Sage Publications. Used with permission.*

the contribution of parameter p_i (with all other parameters fixed) to the variance of ${}^R\hat{x}^*$, represented as $\text{var}({}^R\hat{x}^*)_i$, is calculated as

$$\text{var}({}^R\hat{x}^*)_i = \left(\left. \frac{\partial\, {}^R\hat{x}^*}{\partial p_i} \right|_{\mathbf{p}} \right)^2 \cdot \text{var}(p_i) \qquad (3.42)$$

The independence of the parameters is readily verified from the covariance of $\mathbf{p}$.

Choosing a test point near the centre of the image plane, the contribution of each parameter to the total variance was calculated from equation (3.42) and the results are shown in the far right column of Table 3.1. Importantly, the errors due to ${}^{R}x$ and ${}^{L}x$ are two orders of magnitude greater than the contribution from the system parameters and encoder value. For comparison, the variance in ${}^{R}\hat{x}^{*}$ measured from the bootstrapping process was 0.0035 pixels^2, which agrees well with the sum of contributions from ${}^{L}x$ and ${}^{R}x$. Finally, it should be noted that the variance of ${}^{R}\hat{x}^{*}$ is about half the variance of ${}^{R}x$, indicating that the optimal reconstruction has a higher precision than a single-camera reconstruction.

These results demonstrate the reliability of the image-based techniques presented in Sections 3.2.5 and 3.3 for estimating the encoder value and calibrating the system parameters in the presence of noisy measurements. Furthermore, the main assumptions in deriving equations (3.25) and (3.26) are now justified: any uncertainty in the system parameters and encoder value can be reasonably ignored for the purpose of validation and reconstruction.

3.6 Discussion

In addition to providing a mechanism for validation, the error distance E^{*} in equation (3.26) could be used to measure the random error of range points. As discussed in Section 3.2.1, the error variance of a 3D reconstruction increases with depth as the reconstruction problem becomes ill-conditioned. This systematic uncertainty can be calculated directly from the reconstruction equations (3.28)-(3.30). In contrast, E^{*} measures the random uncertainty due to sensor noise. A suitable function of these systematic and random components could be formulated to provide a unique confidence interval for each 3D point, which would be useful in subsequent processing. For example, parametric surface fitting could be optimized with respect to measurement error by weighting each point with the confidence value.

One of the main limitations of light stripe scanning (compared to methods such as passive stereo) is the acquisition rate. In the current implementation, the PAL frame-rate of 25 Hz results in a 15 second measurement cycle to capture a complete half-PAL resolution range map of 384 stripe profiles. Clearly, such a long acquisition time renders the sensor unsuitable for dynamic scenes. However, a more subtle issue is that the robot must remain stationary during a scan to ensure accurate registration of the measured profiles. Obviously, the acquisition rate could be improved using high-speed cameras and dedicated image processing hardware; high-speed CMOS cameras are now capable of frame-rates exceeding 1000 Hz. Assuming the image processing could be accelerated to match this speed, the sensor could be capable of acquiring 2-3 range maps per second. An example of a high-speed monocular light stripe sensor using a "smart" retina is described by [47].

To minimize complexity and cost, the experimental prototype uses a red laser diode to generate the light plane. Consequently, the scanner only senses surfaces which contain a high component of red. Black, blue and green surfaces reflect insufficient red laser light and are effectively invisible to the sensor. Since the light plane

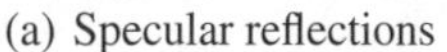

(a) Specular reflections

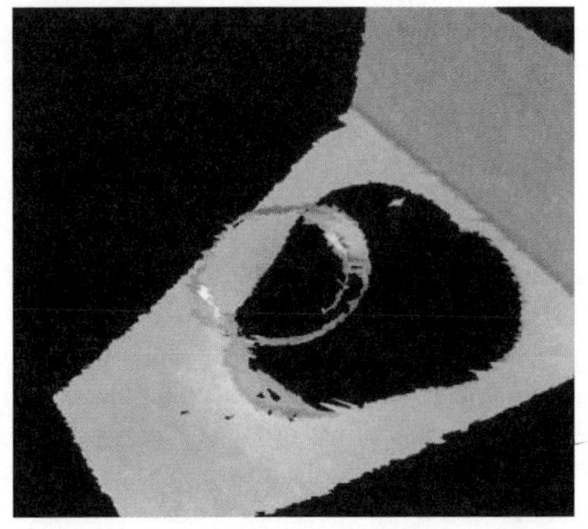

(b) Raw colour/range map using robust scanner

Fig. 3.14. Light stripe scan of a highly polished object. *Reprinted from [151]. ©2004, Sage Publications. Used with permission.*

is not required to be coherent or monochromatic, the laser diode could be replaced by a white light source such as a collimated incandescent bulb. However, laser diodes have particular design advantages including physical compactness, low power consumption and heat generation, and are thus more desirable than other light sources. To solve the colour deficiency problem while retaining these advantages, the light plane could be generated using a triplet of red, green and blue laser diodes. Currently, the main obstacle to this approach is the high cost of green and blue laser diodes.

As with colour, surfaces with high specular and low Lambertian reflection may appear invisible, since insufficient light is reflected back to the sensor. This limitation is common to all active light sensors and can also defeat passive stereopsis, since the surface differs in appearance at each viewpoint. To illustrate this effect, Figure 3.14 shows the raw image and resulting scan of a highly polished object. The only visible regions appearing in the range map are the high curvature edges that provide a specular reflection directly back to the sensor. The best that can be achieved is to ensure that secondary reflections do not interfere with range data acquisition, as demonstrated in this result.

The stripe validation method developed in this chapter may provide an interesting future research direction for *multi-stripe* scanners. Multi-stripe scanners have the potential to solve a number of issues associated with single-stripe scanners: illuminating a target with two stripes could double the acquisition rate, and projecting the stripes from different positions reveals points that would otherwise be hidden in shadow. Current multi-stripe systems rely on different colours or sequences of illumination to disambiguate the stripes (see for example [69]). However, extending the theoretical developments in this chapter could allow multiple stripes from a single scanner to be uniquely identified using the same principles that provide validation for a single stripe.

3.7 Summary and Conclusions

This chapter presented the theoretical framework and implementation of a robust light stripe scanner for a domestic robot, capable of measuring natural scenes in ambient indoor light. The scanner uses the light plane orientation and stereo camera measurements to robustly identify the primary reflection of the stripe in the presence of secondary reflections, cross-talk and other sources of interference. The validation and reconstruction framework is based on minimization of an error distance measured on the image planes. Unlike previous stereo scanners, this formulation is optimal with respect to the measurement error. An image-based procedure for calibrating the parameters of the system from the scan of an arbitrary non-planar target is also demonstrated.

Results from the experimental scanner demonstrate that the proposed method is more effective at recovering range data in the presence of reflections and cross-talk than comparable light stripe methods. Experimental results also confirm the assumptions of the noise model, and show that image-based calibration produces reliable results in the presence of noisy image plane measurements. Finally, the optimal reconstructions from the proposed robust scanner are shown to be more precise than the reconstructions from a single-camera scanner.

The light stripe scanner proposed in this chapter is suitable for operation in an unstructured environment on a service robot. However, the acquisition of dense colour and range data is only the first step towards task planning for autonomous manipulations. The range data must now be interpreted to locate and parameterize target objects for a given task, which is the topic of the next chapter.

4

3D Object Modelling and Classification

Locating and classifying unknown objects is a challenging problem in machine perception, but a crucial skill for performing robotic manipulation tasks flexibly and autonomously. This is particularly evident in domestic applications, where unknown objects are abundant and instances of the same class (cups, bottles and books, for example) can vary significantly in size, colour and shape. In a typical task, the object of interest may vary from a specific instance (ie. your favourite mug) to a more general class (ie. a cup). Thus, a suitable framework must be developed that includes classification of unknown objects in addition to learning and recognizing specific instances.

Many object modelling and recognition techniques are used in practice, and some of the major approaches were briefly outlined in Section 2.3.2. Object representations varying from simple view-based templates to abstract colour histograms, each with corresponding learning and recognition algorithms. The models employed in this chapter are based on simple 3D geometric primitives such as planes, cylinders, cones and spheres. The following sections demonstrate how to automatically and robustly extract these geometric primitives from captured range data (see Chapter 3). Many domestic objects are readily modelled and recognized in terms of such primitives, and recognition is enhanced by augmenting the geometry with colour and texture. Furthermore, primitives automatically extracted from range data can be used to classify unknown objects.

The algorithm developed in this chapter first segments range data into regions modelled by geometric primitives, then constructs an attributed adjacency graph (nodes representing primitives and edges describing adjacency) to describe the relationship between primitives. Predefined object classes are also represented by adjacency graphs, and unknown objects in the scene are recognized by applying sub-graph matching. Specific objects can be identified by examining nodal attributes such as size and colour. A critical component of this algorithm is the robust, non-parametric *surface type* classifier that provides an initial range data segmentation. Experimental results demonstrate that this new classifier produces less over-segmentation in the presence of noise than conventional classifiers, without additional computational expense. Reduced over-segmentation produces larger initial segments that can be robustly fitted with data-driven geometric primitives.

The following section outlines the motivation for the object modelling and classification framework. Section 4.2 introduces the Gaussian image, which is a central concept in surface type classification. Section 4.3 provides an overview of the processing steps in the segmentation algorithm, followed by the technical details of surface type classification and fitting geometric primitives in Sections 4.4 and 4.5. Following segmentation, Section 4.6 describes the construction of adjacency graphs to classify unknown objects. The basic principles are demonstrated with several simple classes including boxes and cups. Finally, Section 4.7 provides an experimental demonstration of the proposed method for a variety of common objects, using real range images. The surface type classification algorithm is also experimentally compared to conventional methods to verify the increased robustness. VRML models of all results can be found in the *Multimedia Extensions*, along with source code and data sets to implement the algorithms.

4.1 Motivation

To achieve maximum flexibility, a domestic robot must deal with the countless objects it will encounter in an efficient manner. This motivates the development of a 3D object recognition framework that can identify instances, which may not have been previously seen, of classes of objects using minimal prior knowledge. The framework must also allow new classes or specific instances of objects to be autonomously learned. The selection of an appropriate object representation is critical to achieving these goals. Representations in previous work on 3D object recognition include CAD models [87], data-driven geometric primitives [167], Generalized Cylinders [23, 126], and non-parametric approaches [108]. Scene interpretation based on CAD models may provide good classification, but does not satisfy the requirement of minimal prior knowledge. At the other extreme, non-parametric interpretations do not attempt to model any part of a scene. This approach limits task planning, as objects cannot be classified or extrapolated into occluded regions.

Classification based on data-driven parametric surfaces offers a suitable compromise for handling unknown objects. Generalized Cylinders is one such approach, in which an object is modelled as a planar cross-sectional function swept about an axis. The cross-section and axis may be arbitrary, but simple functions are chosen in practice to minimize computational expense. This approach has been used successfully for scene interpretation in robotic applications, using both range data [126] and image-based measurements [23]. However, Generalized Cylinders are unsuitable for most tasks since the simplified shapes used in practice do not provide sufficient detail for classifying diverse objects. Based on the above considerations, geometric primitives (planes, spheres, cylinders and cones) are chosen as the preferred framework for data-driven modelling in this research. Geometric primitives are relatively inexpensive to compute, provide distinct classes to aid classification, and faithfully describe the appearance of many simple domestic objects.

A variety of range data segmentation algorithms using parametric surface models have been presented in previous work, and are typically based on a few simple principles. The most common region growing approach involves growing a set of seed regions by iteratively adding range elements that satisfy a hypothesized surface model. Typical surface models include geometric primitives [100] and variable-order bivariate polynomials [12]. The complementary approach assumes range elements are initially connected, and splits the scene into regions at depth discontinuities, local extremes in curvature [35] and changes in local surface shape (also known as *surface type*) [50, 154]. Split-and-merge techniques allow regions to merge after the initial segmentation or split further to improve global consistency [44]. A comparison of prior work in range data segmentation can be found in [62].

The segmentation algorithm presented in this chapter requires only a single split and merge step, by exploiting the increased robustness achieved by fitting primitives to larger initial segments. The algorithm attempts to maintain large segments by initially partitioning range data at jump boundaries and creases, and then further at *surface type* boundaries if the initial segments are not adequately modelled. Subsequent merging of over-segmented regions is based on an iterative boundary cost minimization scheme [44]. While the steps in the segmentation algorithm are not new, the main novelty of this work is the surface type classifier that achieves greater robustness to noise than existing methods, and significantly reduces the degree of initial over-segmentation.

Surface type classification is the process of assigning a local shape descriptor to each range element, and is a central component of the segmentation algorithm. Conventional classification is based on estimating the mean and Gaussian curvatures by fitting parametric surfaces to the range data [12, 50, 154]. However, this process is sensitive to measurement noise, and the arbitrary selection of an approximating surface can introduce systematic errors [90]. This book makes an important contribution to surface type analysis by demonstrating that robust classification can be performed without the need for parametric surfaces. The proposed classifier is based on the analysis of principal curvatures (from the Gaussian image [64]) and surface convexity of small patches of range data. Six surface types are distinguished with greater robustness to noise than conventional classifiers, and without additional computational expense. Thus, the proposed classifier simplifies segmentation by producing larger initial segments of homogeneous surface type.

4.2 The Gaussian Image in Computer Vision

Figure 4.1 describes the relationship between the surface representation and the corresponding Gaussian image of a cylinder. A normal vector can be calculated at each point on the surface by applying a local planar approximation. The Gaussian image is then formed by discarding the spatial information and plotting the normal vectors on a unit sphere (also known as a *Gaussian sphere*). The *extended* Gaussian image (EGI) is a useful representation for polyhedra, and is constructed by associating with each normal vector a scalar value proportional to the area of the corresponding

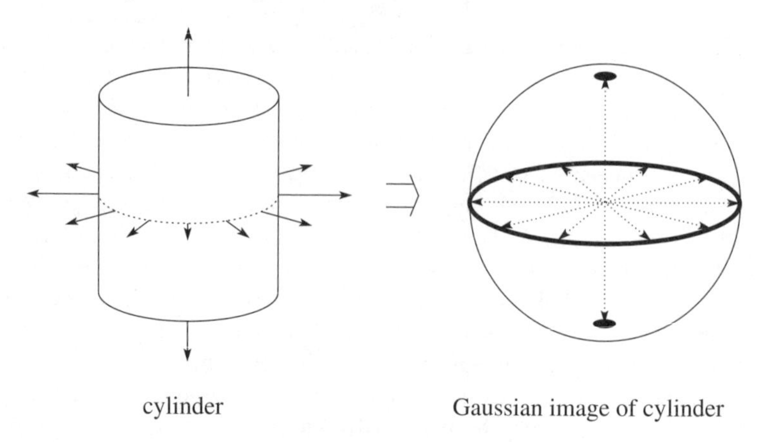

Fig. 4.1. The Gaussian image of a cylinder.

face. Kang and Ikeuchi [79] extend the concept even further with the introduction of the complex extended Gaussian image, which encodes the area, position and normal direction as a complex value on the Gaussian sphere.

In computer vision research, Gaussian images have found application in common tasks including object recognition [120], localization [79] and segmentation [28]. The usefulness of the EGI arises from the property that it is more compact than the associated surface model but nevertheless provides a unique representation for convex polyhedra, ie. the mapping between surface model and Gaussian image is invertible [64]. Often, computer vision algorithms do not use the EGI directly, but instead compute a discrete approximation called an orientation histogram. The Gaussian sphere is tessellated into a finite number of bins, and each records the number of range elements with a normal vector falling within its bounds. Various tessellations are employed, including both uniform [143] and non-uniform [111] schemes.

For object recognition and localization, the Gaussian image is used as a complete template for matching entire objects [79, 120]. Other computer vision tasks exploit direct analysis of features in the Gaussian image. For example, Okatani and Deguchi [121] relate the sign of the Gaussian curvature to 'folds' in the Gaussian image, while Sun and Sherrah [143] detect planes of symmetry by exploiting the association between the symmetry of an object and its Gaussian image. Nayar *et al.* [111] extract features such as moments of inertia, homogeneity and polygonality from an orientation histogram to assess the quality of solder joints (a similar scheme was revisited by Ryu and Cho [134]). In this case, the analysis was simplified by the known relationship between the sensor and target surface.

The surface type classifier proposed in this chapter exploits the observation that the Gaussian image for different shapes exhibit compact and unique structures. Figure 4.1 demonstrates how planes reduce to local maxima on the Gaussian image, while cylinders map to great circles. This property was exploited for range data segmentation in the early work of Dane and Bajcsy [28]. Their analysis involved clustering cells in the orientation histogram, then examining the spread of points in a selected cluster to determine if the associated surface was planar or quadric. However, the results were found to vary significantly with the resolution of the orientation histogram. Vanden Wyngaerd and Van Gool [166] presented a similar analysis of the orientation histogram to find planes, cylinders and cones in range data. The disadvantage of their clustering approach is that it cannot distinguish spheres, which do not form peaks in the Gaussian image. An alternative framework was presented by Chaperon and Goulette [19] to identify cylinders in range maps. The analysis is based on fitting a plane to random sets of points in the Gaussian image, and evaluating a cost function to assess if the plane passes though a great circle of points. However, random sampling is sub-optimal and occasionally leads to gross errors. The classifier proposed in this chapter searches for similar structures in the Gaussian image, and can identify spheres in addition to planes and cylinders. Furthermore, orientation histograms are not required.

Cohen and Rimey [26] developed a segmentation algorithm using surface type classification from Gaussian images, and is similar in philosophy to the method proposed in this chapter. Local patches of normal vectors are classified as planar, cylindrical or spherical using an iterative maximum likelihood scheme, and pixels are clustered into regions according to surface type. In contrast, this chapter presents a closed-form solution to the classification problem, and includes analysis of convexity to distinguish six surface types instead of only three.

4.3 Segmentation Algorithm

We now outline the segmentation algorithm for extracting geometric primitives from range data, which is based on a conventional split then merge approach. The input is a range image of 3D points $\mathbf{M}_i$. The orientation of each surface element is represented by $\mathbf{N}_i$, and the set of points in the $M \times M$ surface patch centred at $\mathbf{M}_i$ are denoted by R_i (typically $M = 15$). The steps in the segmentation algorithm are outlined below:

1. Approximate surface normals $\mathbf{N}_i$ are calculated for each element by taking the cross product of average surface tangents over rows and columns of the surface patch R_i. This method is faster than the usual plane fitting approach and sufficiently accurate for the purpose of segmentation.
2. Surface patches are classified as spanning a depth discontinuity when the average point $\bar{\mathbf{M}}_i$ is sufficiently far from the centre:

$$|\bar{\mathbf{M}}_i - \mathbf{M}_i| > d_M, \quad \bar{\mathbf{M}}_i = \frac{1}{M^2} \sum_{j \in R_i} \mathbf{M}_j \tag{4.1}$$

for a fixed threshold d_M. Patches are classified as spanning a crease when the maximum magnitude of the difference between the central normal and another normal in the patch exceeds a threshold d_N:

$$\max_{j \in R_i} |\mathbf{N}_j - \mathbf{N}_i| > d_N \tag{4.2}$$

Points near discontinuities and creases are removed from the range map.

3. Binary connectivity is applied to the remaining elements to identify smoothly connected regions. Using the methods described later in Section 4.5, the algorithm attempts to fit a geometric primitive to each region, selecting the model with the lowest residual error as the best description.
4. Surface normals are recalculated for each patch R_i using only data from within the same smoothly connected region to reduce perturbations that arise at surface discontinuities. Using the algorithm in Section 4.4, each element is then classified into one of the six surface types illustrated in Figure 4.2, by analysing the Gaussian image and convexity of normal vectors in the patch.
5. Each region is checked for consistency between the fitted model and dominant surface type. For example, regions fitted as cylinders should have a high proportion of surface elements classified as ridges or valleys. If the fitted primitive does not agree with a sufficient proportion of the surface type for elements in the same region (less than 70% in the current implementation), the region is split into sub-regions of homogeneous surface type. The algorithm then attempts to fit a geometric primitive to sub-regions larger than a threshold size. If a region cannot be modelled after splitting by surface type, the constituent pixels are labelled as unclassified and allowed to join other regions in the following step.
6. Iterative region growing assigns unlabelled range elements to existing regions. Regions are iteratively grown by adding neighbouring pixels that satisfy the constraints of the surface model to within a threshold error. After region growing, normal vectors are recalculated and geometric primitives are re-fitted to each region, providing an opportunity for models to change if required.
7. Finally, regions are merged using iterative boundary cost minimization [44]. For each *pair* of adjacent regions, a geometric model is fitted using the *combined* range points. At each iteration, the pair with the lowest combined residue are merged, and the combined models for all neighbouring regions are updated. Merging continues until the minimum combined residue for all neighbouring regions exceeds a threshold.

4.4 Non-parametric Surface Type Classification

Surface type classification is used to verify fitted primitives and identify homogeneous regions when depth discontinuities and creases are insufficient for segmentation. Six of the eight fundamental surface types defined in [12] are considered here, as illustrated in Figure 4.2. As noted by Trucco and Fisher [154], the omitted surface types (*saddle ridge* and *saddle valley*) are not perceptually significant. Classification is based on two properties of the local surface patch R_i: convexity, and the number of non-zero principal curvatures (determined from the Gaussian image).

The principal curvatures of R_i are defined as the maximum and minimum curvatures of a line formed by the intersection of the surface patch and a plane parallel to $\mathbf{N}_i$ and passing through the central point $\mathbf{M}_i$. By inspection, both principle curvatures of a planar patch are zero, a cylindrical patch (ridge/valley) has one zero

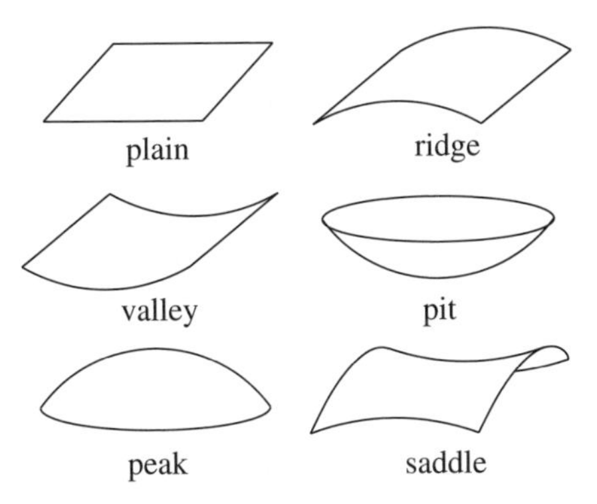

Fig. 4.2. Six surfaces types classified by local shape (the *saddle ridge* and *saddle valley* defined in [12] are not considered in this chapter). *Reprinted from [148]. ©2003, IASTED. Used with permission.*

and one non-zero principal curvature, and a saddle or spherical patch (pit/peak) has two non-zero principal curvatures. Thus, patches can be classified as planar, cylindrical or spherical/saddle by determining the number of non-zero principal curvatures. Section 4.4.1 shows how the number of non-zero principal curvatures can be robustly determined from the Gaussian image of R_i. The distinction between the three groups defined by principal curvatures and the six classes shown in Figure 4.2 depends only on convexity: ridges/peaks are convex, valleys/pits are concave, and planes/saddles are neither. Section 4.4.2 describes how to robustly measure convexity from the spatial distribution of normal vectors.

For segmentation, the convexity and number of principal curvatures are combined to assign a surface type label to each patch. Table 4.1 summarizes the classification rules and the conditions satisfied by each surface type. The main difference between this method and conventional surface type analysis (based on the sign of the mean and Gaussian curvature) is the manner in which planes and saddles are treated. The mean curvature H of a patch is related to the principal curvatures, κ_1 and κ_2, by $H = \frac{1}{2}(\kappa_1 + \kappa_2)$. Thus, the sign of the mean curvature is equivalent to convexity: pits and valleys are concave ($H < 0$), peaks and ridges are convex ($H > 0$), while planes and saddles are neither ($H = 0$). The Gaussian curvature K is related to the principle curvatures by $K = \kappa_1 \kappa_2$. Thus, conventional classification places planes in the same super-class as ridges and valleys ($K = 0$), while saddles form an independent class ($K < 0$). Conversely, classification based on the number of principal curvatures groups saddles with pits and peaks (two non-zero principle curvatures) while assigning planes to an independent class.

4.4.1 Principal Curvatures

The Gaussian image of a surface patch R_i is formed by plotting the normal vectors on the Gaussian sphere. Figure 4.3 plots the Gaussian images for planar, cylindrical

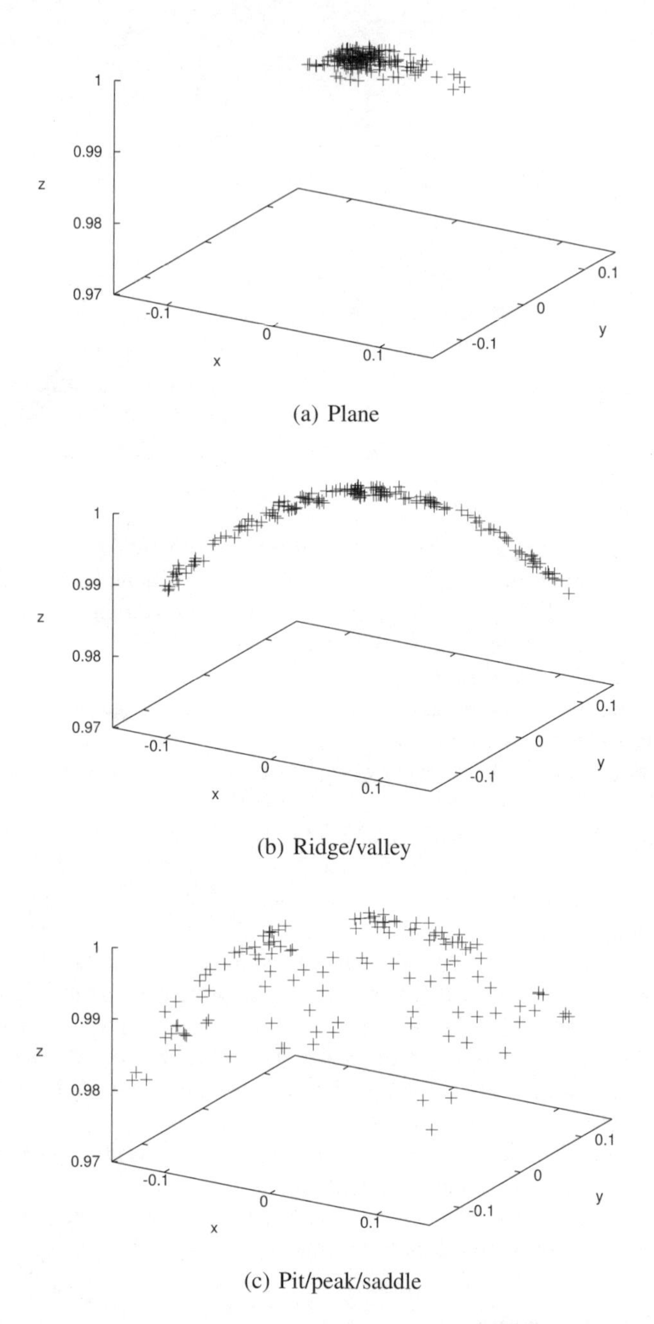

Fig. 4.3. Gaussian images for characteristic surface shapes from experimental data. *Reprinted from [148]. ©2003, IASTED. Used with permission.*

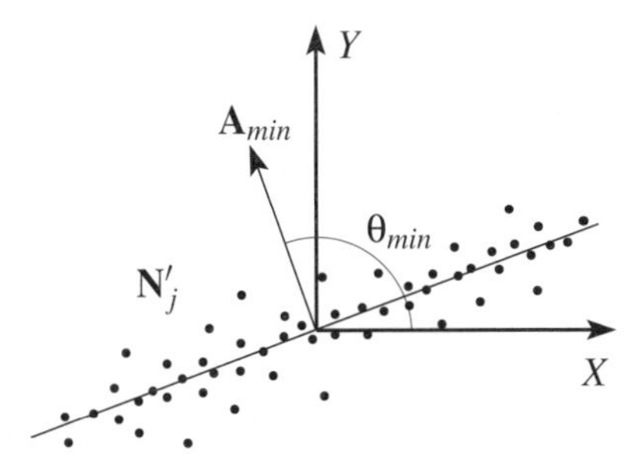

Fig. 4.4. Measurement of normal vector distribution for a cylinder ($\mathbf{A}_{min}$ is a vector in the direction of minimum spread). *Reprinted from [148]. ©2003, IASTED. Used with permission.*

and spherical surface patches obtained from experimental data. Clearly, the planar normals are clustered in the same direction, the cylindrical normals are distributed over a small arc and the spherical normals exhibit a uniform spread. These observations reveal that the number of non-zero principal curvatures can be estimated by measuring the maximum and minimum spread of normal vectors in the Gaussian image. We now show how this spread can be determined from the residue of a fitted plane constrained to pass through the centre of the range patch.

Let $N_p \in [0,1,2]$ represent the number of non-zero principal curvatures and $\mathbf{N}_j$, $j \in R_i$, represent the set of normal vectors in the surface patch R_i, with the central normal denoted as $\mathbf{N}_i$. To simplify the analysis, the normal vectors $\mathbf{N}_j$ are rotated to align $\mathbf{N}_i$ with the Z-axis $\widehat{\mathbf{Z}}$. This transformation can be written as

$$\mathbf{N}'_j = \mathrm{R}[\cos^{-1}(\mathbf{N}_i^\top \widehat{\mathbf{Z}}), \mathbf{N}_i \times \widehat{\mathbf{Z}},] \cdot \mathbf{N}_j \tag{4.3}$$

where $\mathrm{R}(\phi, \mathbf{A})$ is the rotation matrix for a rotation of angle ϕ about axis $\mathbf{A}$. Figure 4.4 illustrates the transformed normal vectors projected onto the XY-plane for the case of a cylindrical patch. Let $\mathbf{A} = (\cos\theta, \sin\theta, 0)^\top$ represent a unit vector in the XY-plane with angle θ to the X-axis. Now, the spread of normal vectors in direction $\mathbf{A}$ can be measured as the residue of a plane with normal $\mathbf{A}$, constrained to pass through the origin (ie. coplanar with $\mathbf{N}_i$), fitted to $\mathbf{N}'_j$. The mean square error in direction $\mathbf{A}$ is:

$$e = \frac{1}{M^2} \sum_{j \in R_i} (\mathbf{N}'^\top_j \mathbf{A})^2 = \frac{1}{M^2} \sum_{j \in R_i} (X_j \cos\theta + Y_j \sin\theta)^2 \tag{4.4}$$

where $\mathbf{N}'_j = (X_j, Y_j, 0)^\top$. The maximum and minimum spread of normal vectors are at the extremum of equation (4.4); Figure 4.4 illustrates the fitted plane and direction of minimum spread (for a cylinder). The optimization is performed in the usual manner by setting $\frac{de}{d\theta} = 0$, which results a quadratic function of $\tan\theta$:

$$\sum_{j \in R_i} (X_j Y_j \tan^2\theta + (X_j^2 - Y_j^2)\tan\theta - X_j Y_j) = 0 \tag{4.5}$$

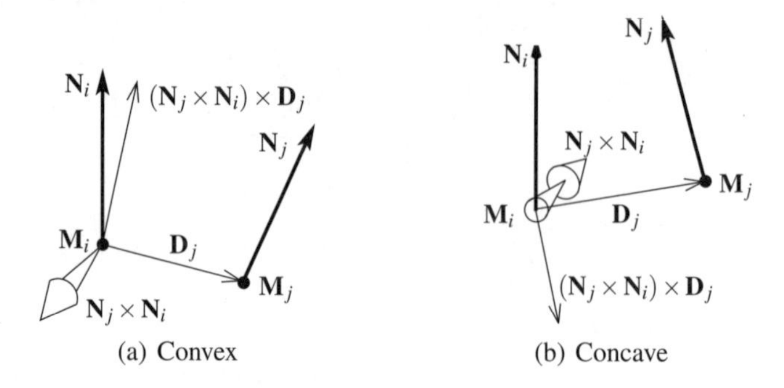

Fig. 4.5. Measuring the convexity between two surface points $\mathbf{M}_i$ and $\mathbf{M}_j$. *Reprinted from [148]. ©2003, IASTED. Used with permission.*

Equation (4.5) has the standard solutions:

$$\theta_{max}, \theta_{min} = \tan^{-1}\left(\frac{\sum_j (Y_j^2 - X_j^2) \pm \sqrt{\Delta}}{2\sum_j (X_j Y_j)}\right) \tag{4.6}$$

where

$$\Delta = [\sum_{j \in R_i} (X_j^2 - Y_j^2)]^2 + 4[\sum_{j \in R_i} (X_j Y_j)]^2 \tag{4.7}$$

The corresponding maximum and minimum residues $e_{max} = e(\theta_{max})$ and $e_{min} = e(\theta_{min})$ are calculated from equation (4.4). When $e_{max} > e_{th}$ or $e_{min} > e_{th}$ for a fixed threshold e_{th}, the corresponding principal curvature is considered to be non-zero. The total number of non-zero principal curvatures N_p is calculated as the number of $e_{min}, e_{max} > e_{th}$, which classifies the patch as a plane, cylinder or sphere/saddle. Selection of the threshold e_{th} depends on the quality of the range data, and can be estimated by observing the spread of normal vectors for a planar surface patch.

4.4.2 Convexity

The convexity of R_i can be determined by observing that the normal vectors in a concave patch tend to converge towards the centre, while the normals in a convex patch tend to point away. Figure 4.5 illustrates a robust measure of convexity based on this principle. For range points in R_i, let $\mathbf{D}_j = \mathbf{M}_j - \mathbf{M}_i$ represent a vector pointing from the central surface element $\mathbf{M}_i$ to a non-central surface element $\mathbf{M}_j$. Then, for each normal vector $\mathbf{N}_j$, the angle between the central normal $\mathbf{N}_i$ and the vector product $(\mathbf{N}_j \times \mathbf{N}_i) \times \mathbf{D}_j$ is greater or less than $\pi/2$ depending on whether $\mathbf{N}_j$ points towards or away from $\mathbf{N}_i$. Each element $j \in R_i$ can therefore be classified as locally convex or concave using the following test:

$$[(\mathbf{N}_j \times \mathbf{N}_i) \times \mathbf{D}_j]^\top \mathbf{N}_i \begin{cases} > 0 : \textit{convex} \\ < 0 : \textit{concave} \end{cases} \tag{4.8}$$

Table 4.1. Surface type classification rules

	Num. Principal Curvatures		
	$N_p = 0$	$N_p = 1$	$N_p = 2$
Convex $S > S_{th}$	Plane	Ridge	Peak
Concave $S < 1/S_{th}$	Plane	Valley	Pit
Neither $1/S_{th} < S < S_{th}$	Plane	Saddle	Saddle

After evaluating the convexity of each element j, the number of locally convex elements N_{cv} and locally concave elements N_{cc} is counted over the whole patch R_i. Now, let $S = N_{cv}/N_{cc}$ represent the ratio of the number of locally convex to locally concave elements. Global patch convexity is determined by the dominant local property: convex when $S > S_{th}$, concave when $S < 1/S_{th}$, or neither when $1/S_{th} < S < S_{th}$, where S_{th} is a fixed threshold ratio. The threshold is ideally unity, but in practice $S_{th} = 1.5$ is used to ensure that the dominant property represents a significant majority. Finally, the statistical measures N_p and S are combined to characterize each patch into one of the six classes shown in Figure 4.2, using the classification rules summarized in Table 4.1.

It should be noted that if the range elements in a patch closely follow one of the classified shapes, the local convexity in equation (4.8) could have been measured using only the difference in range points $\mathbf{D}_j$ and non-central normal $\mathbf{N}_j$ with the simplified condition:

$$\mathbf{N}_j^\top \mathbf{D}_j \begin{cases} > 0 : \mathit{convex} \\ < 0 : \mathit{concave} \end{cases} \tag{4.9}$$

However, the condition given by equation (4.8) was found to be more accurate in the presence of range data noise without excessive computational expense.

4.5 Fitting Geometric Primitives

Geometric primitives are fitted to range segments during three stages of the segmentation algorithm: after the initial segmentation due to creases and depth discontinuities, after growing regions with unlabelled range elements, and finally when determining whether a pair of segments should be merged. Each time a geometric model is required, the algorithm fits all possible primitives to the range data and selects the description with the minimum mean square error. Model fitting is thus an important and computationally expensive component of the algorithm and requires an efficient, robust solution. The following sections describe techniques that can be used to fit planes, spheres, cylinders and cones, including novel estimates of the initial model parameters for passing to a numerical least squares solver.

4.5.1 Planes

Planes are parameterized by a normal vector **N** and a perpendicular distance to the origin d, determined using least squares regression. For a segmented region with N position vectors $\mathbf{M}_i$, $i = 1 \dots N$, the normal vector **N** is the eigenvector associated with the smallest absolute eigenvalue λ of the covariance matrix Λ for $\mathbf{m}_i$ [36], which is given by

$$\Lambda = \frac{1}{N}\sum_{i=0}^{N}(\mathbf{M}_i - \bar{\mathbf{M}})(\mathbf{M}_i - \bar{\mathbf{M}})^\top, \quad \bar{\mathbf{M}} = \sum_{i=0}^{N}\mathbf{M}_i \tag{4.10}$$

The perpendicular distance to the origin is calculated from the mean position $\bar{\mathbf{M}}$ as $d = -\mathbf{N}\cdot\bar{\mathbf{M}}$. The mean square regression error is the minimum eigenvalue λ.

4.5.2 Spheres

Spheres are parameterized by a radius r and centre **C**, which are estimated by minimization of the mean square distance e_{sph}^2 of elements $\mathbf{M}_i$ from the estimated surface:

$$e_{sph}^2 = \frac{1}{N}\sum_i(|\mathbf{M}_i - \mathbf{C}| - r)^2 \tag{4.11}$$

The optimization is performed using Levenberg-Marquardt (LM) minimization, as implemented in MINPACK [105]. Fast convergence of the algorithm relies on accurate initial estimates of the model parameters, denoted as r_0 and $\mathbf{C}_0$, and a novel method for robustly estimating these values is now introduced. Ideally, the centre of the sphere is located at a distance of r in the direction of the surface normal $\mathbf{N}_i$ from the corresponding surface point $\mathbf{M}_i$. Assuming $\mathbf{N}_i$ are uniformly directed towards or away from **C** and taking the mean over all samples, the centre of the sphere $\mathbf{C}_0$ for a given radius r_0 is estimated as:

$$\mathbf{C}_0 = \frac{1}{N}\sum_i(\mathbf{M}_i + r_0\mathbf{N}_i) \tag{4.12}$$

The error variance in the estimated $\mathbf{C}_0$ is:

$$\sigma_{\mathbf{C}}^2 = \frac{1}{N}\sum_i(\mathbf{M}_i + r_0\mathbf{N}_i - \mathbf{C}_0)^\top(\mathbf{M}_i + r_0\mathbf{N}_i - \mathbf{C}_0) \tag{4.13}$$

The initial radius r_0 is calculated as the value that minimizes the variance in the estimate of $\mathbf{C}_0$. Setting $\frac{d\sigma_{\mathbf{C}}^2}{dr_0} = 0$ gives:

$$r_0 = -\frac{N\Sigma_i\mathbf{M}_i^\top\mathbf{N}_i - \Sigma_i\mathbf{M}_i^\top\Sigma_i\mathbf{N}_i}{N\Sigma_i\mathbf{N}_i^\top\mathbf{N}_i - \Sigma_i\mathbf{N}_i^\top\Sigma_i\mathbf{N}_i} \tag{4.14}$$

The initial estimate of $\mathbf{C}_0$ is obtained by substituting equation (4.14) into equation (4.12). Finally, $|r_0|$ and $\mathbf{C}_0$ provide initial values for the numerical minimization of equation (4.11).

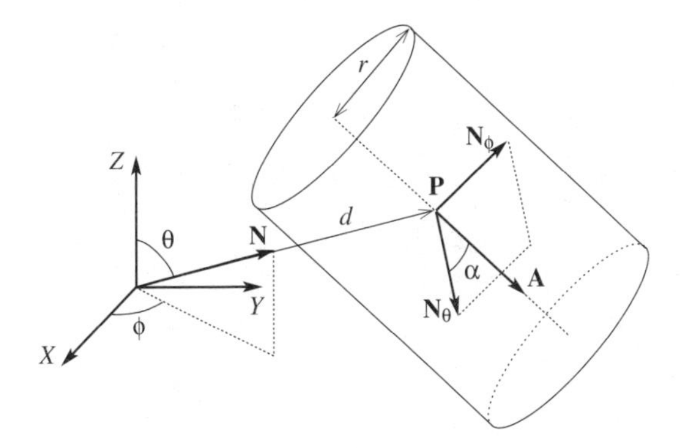

Fig. 4.6. Parameterization of a cylinder.

4.5.3 Cylinders

Cylinders are parameterized using the principles developed in [100], and illustrated in Figure 4.6. Let **P** represent the point on the axis of the cylinder nearest to the origin, which can be represented as a distance d and direction **N** in polar coordinates (θ, ϕ):

$$\mathbf{P} = d\mathbf{N}, \quad \mathbf{N} = (\cos\phi\sin\theta,\ \sin\phi\sin\theta,\ \cos\theta)^\top \tag{4.15}$$

Let **A** represent a unit vector in the direction of the cylinder axis, noting that **A** and **N** are orthogonal. Two basis vectors, $\mathbf{N}_\theta$ and $\mathbf{N}_\phi$, which are orthogonal to each other and **N**, are constructed by taking the partial derivatives of **N** and normalizing to unit length:

$$\mathbf{N}_\theta = (\cos\phi\cos\theta,\ \sin\phi\cos\theta,\ -\sin\theta)^\top \tag{4.16}$$

$$\mathbf{N}_\phi = (-\sin\phi,\ \cos\phi,\ 0)^\top \tag{4.17}$$

The axis **A** can now be expressed as a function of $\mathbf{N}_\theta$, $\mathbf{N}_\phi$ and a single parameter α:

$$\mathbf{A} = \mathbf{N}_\theta \cos\alpha + \mathbf{N}_\phi \sin\alpha \tag{4.18}$$

The complete parameterization of the cylinder includes d, θ, ϕ, α and radius r. As before, the LM algorithm is used to estimate the parameters by minimizing the mean square distance e_{cyl}^2 of range points $\mathbf{M}_i$ from the surface of the cylinder. The error function is:

$$e_{cyl}^2 = \frac{1}{N}\sum_i (|\mathbf{D}_i| - r)^2, \quad \mathbf{D}_i = \mathbf{P} + \mathbf{A}(\mathbf{M}_i - \mathbf{P})^\top \mathbf{A} - \mathbf{M}_i \tag{4.19}$$

where **P** and **A** are given by equations (4.15) and (4.18).

As before, rapid convergence of the model requires accurate initial estimates of the parameters. An initial estimate of the cylinder axis $\mathbf{A}_0$ is obtained using an extension of the surface classification algorithm in Section 4.4. First, the normal

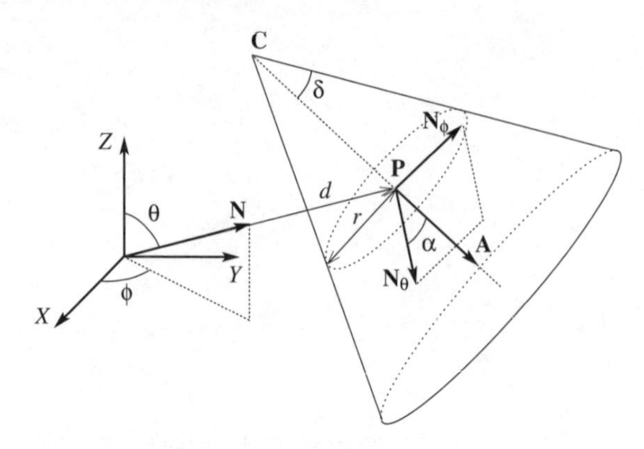

Fig. 4.7. Parameterization of a cone.

vectors are transformed as $\mathbf{N}_i' = \mathrm{R}[\cos^{-1}(\bar{\mathbf{N}}^\top \widehat{\mathbf{Z}}),\ \bar{\mathbf{N}} \times \widehat{\mathbf{Z}}]\mathbf{N}_i$ to align the mean normal with the Z-axis, where the mean normal is given by

$$\bar{\mathbf{N}} = \frac{\mathbf{N}'}{|\mathbf{N}'|}, \quad \mathbf{N}' = \sum_{i=1}^{N} \mathbf{N}_i \tag{4.20}$$

Using equations (4.6)-(4.7), the transformed cylinder axis $\mathbf{A}_0'$ is approximated as the vector in the XY-plane that minimizes the spread of normals in the Gaussian image, as shown by $\mathbf{A}_{min}$ in Figure 4.4. The cylinder axis $\mathbf{A}_0$ is recovered by taking the inverse transformation $\mathbf{A}_0 = \mathrm{R}^{-1}[\cos^{-1}(\bar{\mathbf{N}}^\top \widehat{\mathbf{Z}}),\ \bar{\mathbf{N}} \times \hat{\mathbf{Z}}]\mathbf{A}_0'$.

The initial estimates r_0 and $\mathbf{P}_0$ are obtained using the method introduced in the Section 4.5.2. The samples $\mathbf{M}_i$ and normals $\mathbf{N}_i$ are rotated to align $\mathbf{A}_0$ with the Z-axis, then projected onto the XY-plane to give the circular cross-section of the cylinder. The centre $\mathbf{C}_0'$ and radius r_0 of the circle are estimated from the projected data using equations (4.12) and (4.14). Finally the centre is transformed back to the original frame as $\mathbf{C}_0$, and the closest point on the axis is recovered as $\mathbf{P}_0 = \mathbf{C}_0 - \mathbf{A}_0 \mathbf{C}_0^\top \mathbf{A}_0$.

4.5.4 Cones

Cones are parameterized in a similar manner to cylinders, as illustrated in Figure 4.7. Let $\mathbf{P}$ represent the point on the axis closest to the origin, and $\mathbf{A}$ represent a unit vector in the direction of the axis. Using equations (4.15)-(4.18), these vectors are again parameterized in terms of d, θ, ϕ and α. Now, if the apex is located at $\mathbf{C}$, and the half-angle of the cone is δ, the radius r at $\mathbf{P}$ along the axis can be calculated as:

$$r = (\mathbf{C}^\top \mathbf{A}) \tan \delta \tag{4.21}$$

Thus, the cone is completely parameterized in terms of d, θ, ϕ, α, r and δ. Again, the LM algorithm is used to estimate the cone parameters by minimizing the mean square distance e_{cone}^2 between the samples $\mathbf{M}_i$ and the surface of the cone. The error function is given by:

$$e_{cone}^2 = \frac{1}{N}\sum_i(|\mathbf{D}_i| - r_i)^2, \tag{4.22}$$

where

$$\mathbf{D}_i = \mathbf{C} + \mathbf{A}(\mathbf{M}_i - \mathbf{C})^\top \mathbf{A} - \mathbf{M}_i$$
$$r_i = [(\mathbf{M}_i - \mathbf{C})^\top \mathbf{A}] \tan\delta$$

To obtain initial estimates for the parameters, it is first noted that a normal vector $\mathbf{N}_i$ on the surface of an ideal cone has angle $\psi = \frac{\pi}{2} - \delta$ to the cone axis $\mathbf{A}$. For a given estimate of the axis $\mathbf{A}$, this angle can be estimated as $\psi_i(\mathbf{A}) = \cos^{-1}(\mathbf{N}_i^\top \mathbf{A})$ for measured normal $\mathbf{N}_i$. Taking the average over all normals, an estimate for ψ_0 and its variance σ_ψ^2 are calculated as:

$$\bar{\psi}_0 = \frac{1}{N}\sum_i \cos^{-1}(\mathbf{N}_i^\top \mathbf{A}) \tag{4.23}$$

$$\sigma_\psi^2 = \frac{1}{N}\sum_i [\cos^{-1}(\mathbf{N}_i^\top \mathbf{A}) - \bar{\psi}]^2 \tag{4.24}$$

The axis can now be estimated as the direction that minimizes the variance in the above estimate of ψ_0. LM minimization is used with $\mathbf{A}$ expressed in polar coordinates and initially set to the direction of the Z-axis. It should be noted that the same method could have been used to estimate the axis of a cylinder, although the closed form solution presented in Section 4.5.3 is preferred. Once the initial estimate $\mathbf{A}_0$ is found, ψ_0 is calculated from equation (4.23) and an initial value for the cone half angle is recovered as $\delta_0 = \frac{\pi}{2} - \psi_0$.

The initial estimates $\mathbf{C}_0$ and r_0 are obtained in a similar manner to the parameters of a cylinder. First, range elements $\mathbf{M}_i$ and normals $\mathbf{N}_i$ are rotated to align $\mathbf{A}_0$ with the Z-axis. Range points $\mathbf{M}_i'$ are then projected away from the apex, tangentially to the surface of the cone, onto the XY-plane. For point $\mathbf{M}_i'$, the direction of projection is $\mathbf{S}_i = \mathrm{R}(\pi - \delta,\ \widehat{\mathbf{Z}} \times \mathbf{N}_i')\widehat{\mathbf{Z}}$, and the resulting point on the XY-plane is $\mathbf{M}_i'' = \mathbf{M}_i' - (m_{iz}'/s_{iz})\mathbf{S}_i$. The centre and radius of the projected arc are then estimated using equations (4.12) and (4.14). Finally, the results are transformed back to the original frame to recover $\mathbf{C}_0$ and r_0.

4.6 Object Modelling and Classification

The purpose of the scene analysis and object modelling algorithms presented in this section are to associate the abstract mathematical descriptions extracted by segmentation with meaningful semantics for high-level task planning. As mentioned earlier, the implementation described in this book is limited to simple convex objects such as boxes and cans to demonstrate the basic principles. To aid the following discussion, the scene shown in Figure 4.8(a) is analyzed as a specific example. In this case, a successful analysis should identify the presence of a box, ball and cup, and generate suitable textured polygonal models of the objects.

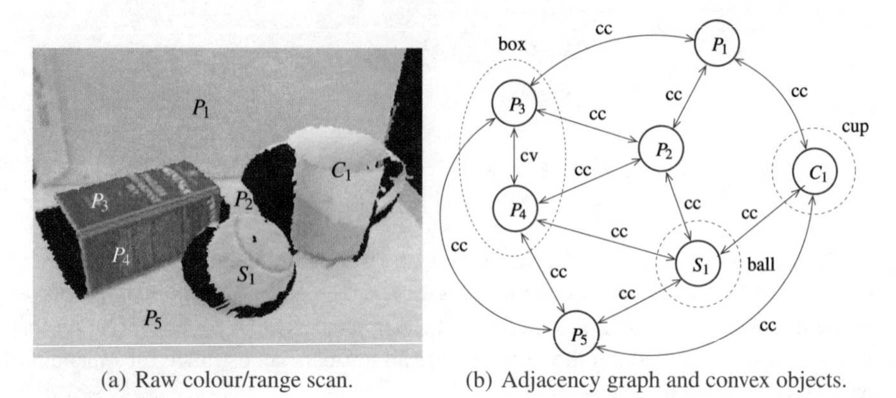

(a) Raw colour/range scan. (b) Adjacency graph and convex objects.

Fig. 4.8. Sample objects for scene analysis and object modelling. The extracted models are shown in Figure 4.11(d).

To recognize unknown objects, the system represents generic classes of objects using attributed adjacency graphs. The nodes of a graph represent geometric primitives, with attributes describing shape parameters and texture information, and edges represent the type of connection (convex or concave) between primitives. Thus, a box is recognized as two or three orthogonal, convexly connected planes (the remaining planes are unobservable), while a can is a cylinder (or cone) convexly connected to a plane. Further constraints (such as the allowed radius of a cylinder) can be placed on the nodal attributes to distinguish similarly shaped objects such as cups and bowls.

The first step in scene analysis is to generate an attributed graph describing the relationship between geometric primitives in the observed scene. Segmentation provides a list of neighbouring regions, and the following tests are applied to determine the type of connection (convex or concave) between adjacent primitives:

Plane/Plane: The convexity of the edge between two planes is determined using the principles described in Section 4.4.2. Let $\mathbf{n}_1$ and $\mathbf{n}_2$ represent the normal vectors of the adjacent planes, and $\mathbf{m}_1$ and $\mathbf{m}_2$ represent the average position of range elements associated with each plane. Then, following the simplified convexity condition in equation (4.9), the convexity of the edge joining the planes is classified according to:

$$\mathbf{N}_1^\top(\mathbf{M}_2 - \mathbf{M}_1)\begin{cases} < 0 : \textit{convex} \\ > 0 : \textit{concave} \end{cases} \tag{4.25}$$

Plane/Cylinder or Plane/Cone: The convexity of an edge between a cylinder (or cone) and a plane is determined by noting that for a closed cylinder, the points on the end-plane lie within the radius of the cylinder. Conversely, the points on the concavely connected planar rim (such as the brim of a hat) lie outside the radius of the cylinder. Thus, the convexity of the connection is classified by measuring the average distance of the N points $\mathbf{M}_i$ in the plane from the axis $\mathbf{A}$

of the cylinder:

$$\frac{1}{N}\sum_{i}|\mathbf{M}_i - [\mathbf{P}+\mathbf{A}(\mathbf{M}_i-\mathbf{P})^\top\mathbf{A}]| \begin{cases} < r : \mathit{convex} \\ > r : \mathit{concave} \end{cases} \tag{4.26}$$

where $\mathbf{P}$ is a point on the axis of the cylinder and r is the radius (average radius in the case of a cone). A similar analysis could be used to determine the convexity of an edge between a plane and a sphere.

For the simplified analysis considered in this chapter, the edges joining all remaining adjacent primitives are classified as concave. Figure 4.8(b) shows the attributed graph for the scene in Figure 4.8(a), where P_i, C_i and S_i denote planes, cylinders and spheres, and cc and cv describe concave and convex connections respectively. Convex objects are identified by removing all concave connections and matching the remaining (convexly connected) sub-graphs to the models for generic classes of objects. Sub-graphs that do not match any class (such as P_1, P_2 and P_5 in Figure 4.8) are ignored. Four classes are identified in the current implementation: boxes, cups/bowls, cans and balls. For the example scene in Figure 4.8(b), the classified sub-graphs are shown by the dotted circles. Finally, a textured polygonal model is generated for each classified object based on the primitives in the sub-graph, as described in the following sections.

4.6.1 Modelling a Box

A rectangular box is modelled as three pairs of parallel planes, with adjacent sides approximately orthogonal. Let A_i and B_i, $i = 1 \ldots 3$, represent the three pairs of parallel planes, where A_i is closest to the observer for each pair. Plane pairs are parameterized by a common normal $\mathbf{N}_i$ and two perpendicular distances a_i and b_i to the origin, where $a_i < b_i$. At most, three sides of the box are visible (planes A_i), and determine the normal vectors $\mathbf{N}_i$ and distances a_i. If only two sides are visible, the normal vector for the third pair is calculated as $\mathbf{N}_3 = \mathbf{N}_1 \times \mathbf{N}_2$, based on the assumption that adjacent faces are orthogonal.

The distance parameters b_i of the hidden faces (including a_3 if only two faces are visible) are then determined by fitting planes to the edges of the visible surfaces using the voting scheme illustrated in Figure 4.9. In this example, b_3 is calculated by extracting the set of range points $\mathbf{E}_j$ along the boundary of the segments associated with adjacent faces A_1 and A_2. The voting scheme assumes that B_3 will be coincident with one of the points $\mathbf{E}_j$ along the adjacent edges. Each edge point $\mathbf{E}_j$ is assigned a likelihood c_j that measures the evidence that B_3 is coincident with $\mathbf{E}_j$. The likelihood is calculated as:

$$c_j = \sum_{k}(|\mathbf{N}_3^\top(\mathbf{E}_k - \mathbf{E}_j)|^2 + 1)^{-1} \tag{4.27}$$

and increases as more edge points are approximately coincident with B_3 passing through $\mathbf{E}_j$. Two local maxima exist, corresponding to the front and rear planes in the pair, and the distance parameter for the hidden plane is calculated as $b_3 = -\mathbf{N}_3^\top\mathbf{E}_{max}$, where $\mathbf{E}_{max}$ is the edge point for the most distant maxima.

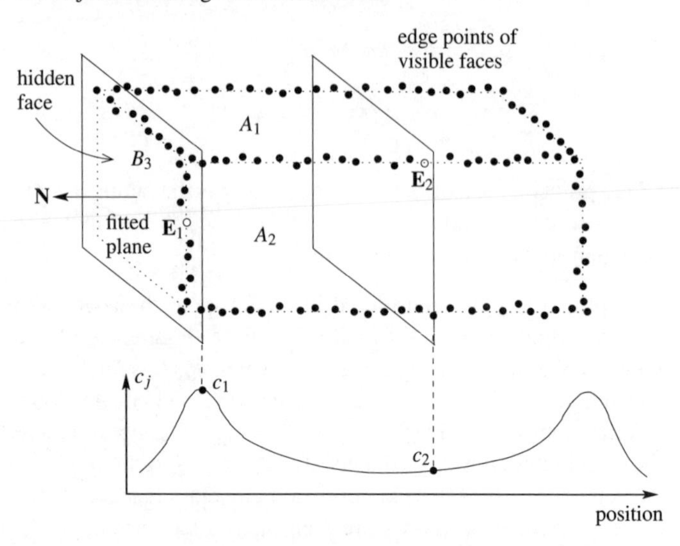

Fig. 4.9. Calculating the position of the hidden face B_3 with known normal $\mathbf{N}_3$ from edge points $\mathbf{E}_j$ (solid black circles) of adjacent faces. The cost function c_j in equation (4.27) gives evidence for the hypothesis that the plane passes through $\mathbf{E}_j$, and has been evaluated for points $\mathbf{E}_1$ and $\mathbf{E}_2$ (hollow black circles). The cost function indicates that $\mathbf{E}_1$ is supported by more coplanar edge points than $\mathbf{E}_2$.

Finally, a polygonal model of the box is constructed. Eight corner vertices at the intersections of the planar faces are found by solving linear systems such as

$$\begin{pmatrix} \mathbf{N}_1^\top \\ \mathbf{N}_2^\top \\ \mathbf{N}_3^\top \end{pmatrix} \mathbf{V}_1 = \begin{pmatrix} a_1 \\ a_2 \\ a_3 \end{pmatrix} \tag{4.28}$$

for the intersection $\mathbf{V}_1$ between A_1, A_2 and A_3. The remaining vertices are calculated similarly, and are used to define the six polygonal faces of the box. The colour/range segment associated with each face is projected onto the surface of the box and rendered into a texture map using conventional computer graphics techniques.

4.6.2 Modelling a Cup/Bowl/Can

Convex cylindrical objects are modelled as cones or cylinders, and zero or more parallel end-planes. Let $\mathbf{A}$ represent the axis of the cylinder or cone, $\mathbf{N}$ represent the common normal of the end-planes, and d_1, d_2 represent the perpendicular distances of the end-planes from the origin, where d_1 corresponds to the upper plane. If either end-plane is visible, the normal direction of the plane is assigned to $\mathbf{N}$, and the distance parameter is assigned to d_1 (or d_2 if the lower plane is visible). Otherwise, the end-plane normal $\mathbf{N}$ is assumed to be parallel to the axis $\mathbf{A}$. The distance parameters d_i for the unobserved end-planes are determined using the voting scheme introduced above. Boundary points are extracted from the range data for the cylindrical region, and the support for a plane with normal **N** passing through each edge point is calculated from (4.27). Again, the likelihood function exhibits two local maxima corresponding to the two end-planes, and the distance d_i to the hidden plane is calculated appropriately.

Finally, a textured polygonal model is constructed to approximate the cylindrical object. Two sets of vertices are generated at fixed angles around the top and bottom rim where the cylinder intersects the end-planes. The vertices are tessellated using triangular facets, and the range/colour data is projected onto the surface of each primitive and rendered into a texture map using conventional computer graphics techniques.

4.6.3 Modelling a Ball

Since a ball is represented by single sphere, constructing an approximate polygonal model is more straightforward than the objects considered above. Intersections between spheres and other geometric primitives are not considered in the current implementation. A polygonal approximation is constructed by parameterizing the surface of the sphere using spherical coordinates, and generating vertices at regular intervals of longitude and latitude. While other tessellations could be used generate a more uniform distribution of vertices, this scheme was found to be sufficient for the experiments considered in here. The range/colour data is finally projected onto the surface of the sphere and rendered into a texture map (in spherical coordinates).

4.7 Experimental Results

The algorithms described in this chapter were tested using experimental range images at half-PAL resolution (384×288 pixels) from the robust stereoscopic light stripe scanner described in Chapter 3. Section 4.7.1 presents the segmentation results for selected scenes containing a variety of domestic objects. Section 4.7.2 then compares the noise robustness of the proposed surface type classifier to the conventional approach, by applying both algorithms to the same range data. VRML models of the results can be found in the *Multimedia Extensions*.

4.7.1 Segmentation and Object Modelling

The first scene in Figure 4.10(a) contains typical domestic objects described by simple geometric primitives: a box, ball and cup. Figure 4.10(b) shows the result after removing depth discontinuities and creases, and calculating the surface type. Range elements with identical surface type are represented in homogeneous colour. The surface labels generally agree with perceptual expectations, although some misclassification is apparent near discontinuities and creases where normal vectors are less reliable. The primitives fitted to regions segmented by discontinuities and creases

(a) Colour/range scan of box, ball and cup.

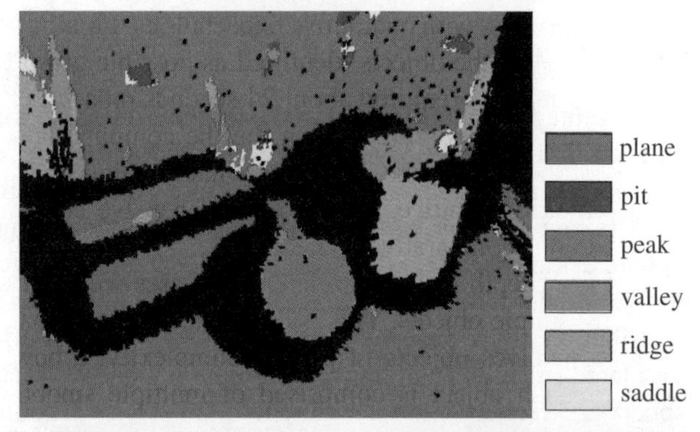

(b) Discontinuity/crease removal and surface type classification.

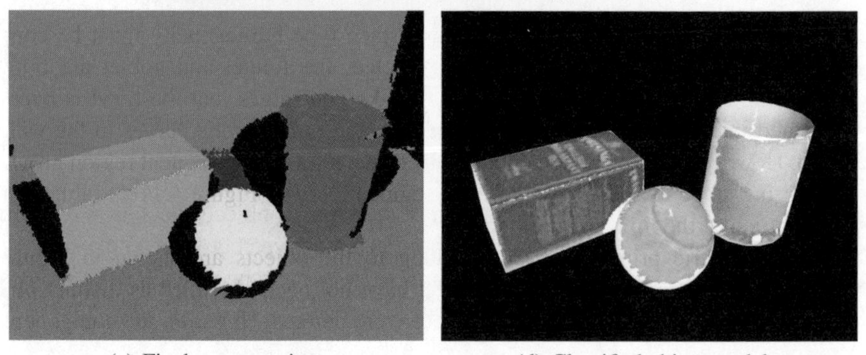

(c) Final segmentation. (d) Classified object models.

Fig. 4.10. Results for box, ball and cup (see also *Multimedia Extensions*).

Table 4.2. Computational expense of main steps in segmentation.

Computation	Time (ms)
normal vector calculation	1730
discontinuity/crease removal	800
surface type labeling	2860
region growing	232
region merging	13200

agree well with the dominant surface type, and further segmentation in this case was not necessary. The final segmentation after growing and merging regions is shown in Figure 4.10(c), and is consistent with qualitative expectations. Finally, Figure 4.10(d) shows a textured rendering of the objects identified as possible grasping targets for the robot. The complete analysis required about 24 seconds on a 2.2 GHz dual Intel Xeon PC, and Table 4.2 provides a breakdown the the computational expense for the main components of the algorithm. Clearly, the final merging stage (which relies heavily on fitting geometric primitives) is the most expensive computation, while surface type classification is comparatively cheap.

In the previous scene, depth discontinuities and creases were sufficient to segment the geometrically simple objects. This is contrasted by the situation shown in Figure 4.11(a), which involves objects of greater complexity: a bowl, funnel and (upside-down) goblet. Each object is comprised of multiple smoothly-connected primitives of different geometry. Figure 4.11(b) shows the result of discontinuity/crease removal and surface type classification. Clearly, discontinuities and creases provide an incomplete segmentation, and surface type homogeneity must be consulted to isolate simple primitives. As expected, the funnel and goblet are both divided into approximately equal areas of peaks and ridges, but the bowl is over-segmented into planes, ridges, valleys and pits due to the subtle curves in the wall and lip. However, this over-segmentation is corrected in the subsequent region growing/merging stage as shown by the final segmentation in Figure 4.11(c) (although segmentation of the lip is still not ideal).

The geometric primitives corresponding to the objects are shown in Figure 4.11(d). In this case, the funnel and goblet have not been classified as distinct objects, since no generic class for these shapes were defined. However, the component primitives agree well with qualitative expectations: the funnel contains a spherical body, conical spout and planar handle, and the cup is decomposed into a planar base, cylindrical stem and a body comprised of a spherical base and cylindrical walls. These results lend support to the possibility of learning new classes based on the observed primitives. Note also that this result (and the result in Figure 4.10(d)) demonstrates the ability to handle partial occlusion. In the first scene, the box was partially obscured by the ball, while both foreground objects masked the funnel in the second. Object modelling was not affected in either case, since sufficient data was available to establish the correct surface model.

(a) Colour/range scan of bowl, funnel and goblet.

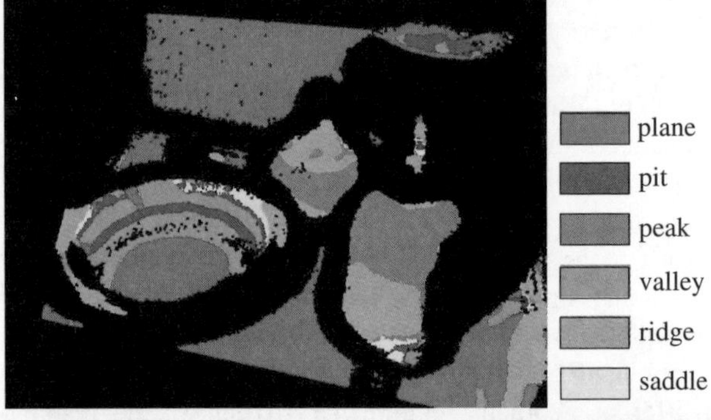

(b) Discontinuity/crease removal and surface type classification.

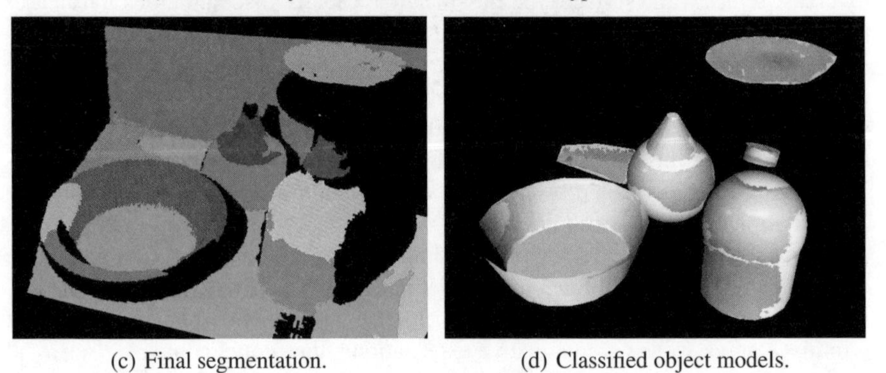

(c) Final segmentation. (d) Classified object models.

Fig. 4.11. Results for bowl, funnel and goblet (see also *Multimedia Extensions*).

(a) Colour/range scan of bottle and bowl.

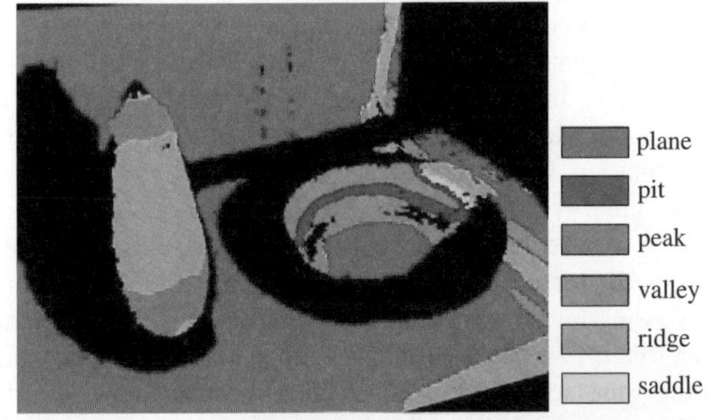

(b) Discontinuity/crease removal and surface type classification.

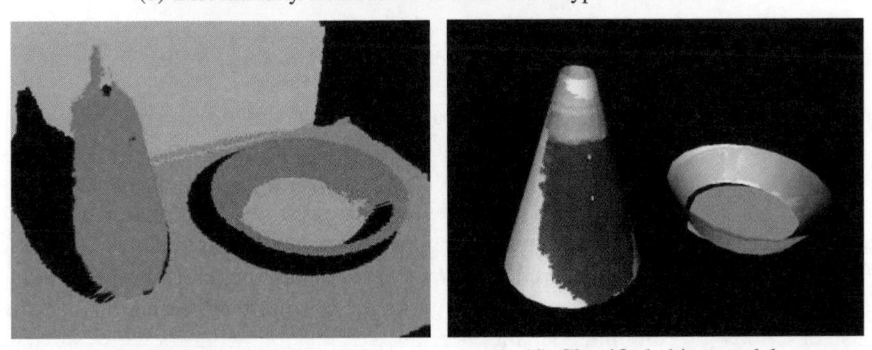

(c) Final segmentation.

(d) Classified object models.

Fig. 4.12. Results for bottle and bowl (see also *Multimedia Extensions*).

Figure 4.12 shows an example of a sauce bottle with subtly curved body that cannot be accurately modelled in terms of simple geometric primitives. Figure 4.12(b) demonstrates the high sensitivity of the surface type classifier to distinguish the conical upper-half of the bottle from the pseudo-spherical lower-half. The final segmentation and model in Figures 4.12 and 4.12 show that a single cone was chosen as the most accurate interpretation of the bottle. Obviously, the sensitivity of the algorithm to subtle variations in surface shape can be adjusted by varying the thresholds for region splitting and merging.

A final example is shown in Figure 4.13(a), and the target in this case is a shallow pyramid approximately 5 cm wide and 3.5 cm high. As before, discontinuities and creases are insufficient for a complete segmentation, but in this case the scene contains only planar surfaces. Figure 4.13(b) shows the outcome of surface type classification, with some interesting results. While the central portion of each face is correctly classified as planar, the edges are perceived as ridges or valleys depending on convexity. Similarly, the corners are labelled as saddle points. On closer inspection, these results accurately reflect the topology of the scene but nevertheless lead to an over-segmentation. Figure 4.13(c) shows the final segmentation after growing and merging the initial regions. In this case, the merging step was unable to completely correct the over-segmentation, and the convex edges between each face are still identified as cylindrical. If this result was used for object recognition, the error might be corrected by introducing heuristics regarding the possible shape of planes intersecting at a shallow angle. Regardless of the segmentation result, Figure 4.13(b) demonstrates the good sensitivity of the proposed surface type classifier to subtle variations in shape.

4.7.2 Curvature-Based Versus Non-parametric Surface Type Classifiers

In this section, the performance of the proposed non-parametric surface type classifier is compared to conventional curvature-based techniques. The surface-type classifier described by Besl and Jain [12] was chosen as representative of this approach. This method performs equally-weighted least squares estimation of first and second order directional derivatives by convolving the range image with window operators. The sign of the mean and Gaussian curvatures are then computed from the derivatives and used to classify each pixel. A method to reduce computational cost is also described, by combining smaller derivative estimation windows with Gaussian smoothing. However, this optimization was not implemented, since the size of the smoothing window was noted to influence the estimated derivatives, for a comparatively small gain in performance.

The curvature-based surface type classifier was applied to the scene shown in Figure 4.10. The range image was computed by taking the magnitude of each 3D point in the range map, and 15×15 pixel window operators were applied to calculate the surface type. From examination of experimental trials, thresholds $\varepsilon_H = 0.005$ and $\varepsilon_K = 0.0001$ were chosen for detection of non-zero mean and Gaussian curvatures. The processing time required by the curvature-based classifier was approximately the same as the non-parametric algorithm, and Figure 4.14 shows the classification

(a) Colour/range scan of pyramid.

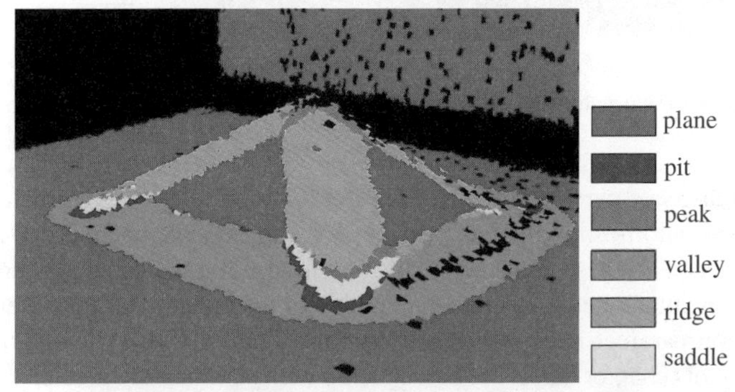

(b) Discontinuity/crease removal and surface type classification.

(c) Final segmentation.

Fig. 4.13. Results for pyramid.

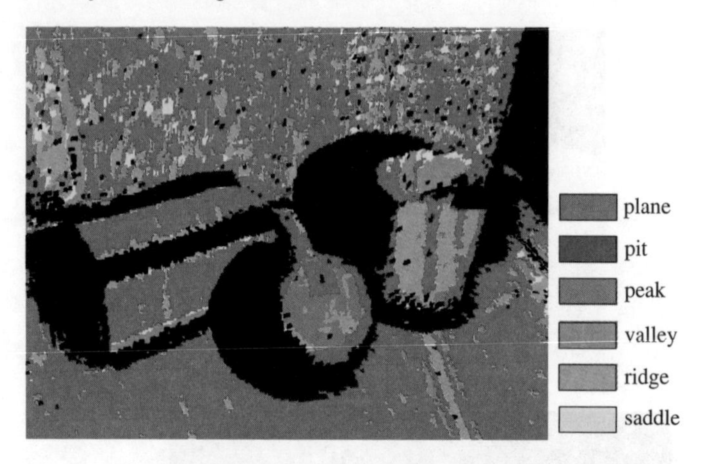

Fig. 4.14. Curvature-based surface type result (compare to Figure 4.10(b)).

result. The areas of particular interest are the ball and interior/exterior faces of the cup, which should be classified as peak, valley and ridge respectively. Detection of the correct shape in these areas is patchy, and modifying the detection thresholds to improve the result tended to degrade other parts of the image. The equivalent analysis using the non-parametric classifier is shown in Figure 4.10(b). Clearly, the method proposed in this chapter labels the ball and cup with significantly greater consistency, and much less noise is evident in the planar floor and wall.

As noted by Besl and Jain, the second order derivative estimates required to calculate mean and Gaussian curvatures result in a high sensitivity to noise. Their proposed segmentation process overcomes this problem by eroding large areas of homogeneous surface type into small (approximately 10 pixel) seed regions, then dilating the seeds using a model-based region growing algorithm. Conversely, the robust labelling produced by the non-parametric classifier proposed in this chapter eliminates the need for erosion/dilation, and produces large regions of homogeneous surface type that can be directly treated as initial segments. Furthermore, the improved robustness comes without additional computational expense.

4.8 Discussion and Conclusions

This chapter presented a new method for surface type classification based on analysis of principal curvatures (from the Gaussian image) and convexity, and utilized the techniques for data-driven object modelling and classification. The proposed non-parametric classifier eliminates the need for choosing arbitrary approximating surface functions as required by methods based on the mean and Gaussian curvature. The new method is comparable in computational expense to existing classifiers, but exhibits significantly greater robustness to noise and produces large regions of homogeneous surface type consistent with perceptual expectations. The proposed classifier is used in a range segmentation algorithm to find geometric primitives in a range map. Common objects are then identified as specific collections of extracted primitives. Experimental results confirm that the proposed technique is capable of successfully classifying and modelling a variety of domestic objects using planes, spheres, cylinders and cones.

The main computational cost of segmentation is the final iterative merging stage. This involves fitting every neighbouring pair of regions with all possible geometric primitives to determine the pair to be merged. Since most pairs will not be merged, a great deal of processing time is simply wasted. It may be possible to accelerate the merging stage by introducing heuristics (based on, for example, analysis of the Gaussian image) to quickly eliminate some pairs from consideration before committing to an expensive fitting algorithm. However, the current processing speed is reasonable for a service robot, particularly if the extracted objects are subsequently tracked using a faster algorithm, which is the topic of the next chapter.

As with many segmentation algorithms, the methods employed here require several empirical thresholds. These include selection of the window size for normal vector and surface type calculations, and thresholds to detect creases, non-zero principal curvatures and acceptable surface models. Parameter selection is driven not only by consideration of the sensor noise, but also by the needs of object classification and robotic grasp planning. For example, cups with planar sides, embossed patterns and curved profiles are common in practice, but can be recognized as generally cylindrical at an appropriate scale. Furthermore, features on the scale of the hand (such as the height and radius of a cup) have a greater impact on grasping than smaller surface variations. The window size for normal vector and surface type calculations determines the smallest detectable features. Similarly, the sensitivity of surface classification is controlled by the threshold for detection of non-zero principal curvatures. The particular choice of parameters in the experimental results was found to segment objects at a scale appropriate for classification and grasp planning. However, for applications such as CAD modelling, it may be appropriate to select parameters which produce finer details.

The ability of the proposed algorithm to model complex objects is clearly limited by the selection of geometric primitives. An obvious direction for further work is to consider additional primitives such as ellipsoids and tori. However, any addition of new primitives must also consider the tradeoff between generality, complexity and computational expense. Alternatively, a greater variety of domestic objects could be classified by developing more sophisticated object modelling algorithms to handle complex relationships between primitives, which would ultimately allow a robot to perform a greater range of tasks. Another interesting research direction would be to exploit the ability of the robot to interact with the world to aid scene understanding. If the initial models provide sufficient information to grasp an object, the robot could then view the object from all sides to refine or verify the initial classification. Our own tendency to explore the world in this way could provide valuable insight into developing similar "active understanding" strategies.

5

Multi-cue 3D Model-Based Object Tracking

Once an object has been located and classified using the techniques in the previous chapters, the system must continue to update the estimated pose for several reasons. Clearly, the initial pose will quickly become invalid if the object is under internal or external dynamic influences. However, even if the object is static, the motion of the active cameras may dynamically bias the estimated pose through modelling errors. Even a small pose bias is sufficient to destabilize a planned grasp and cause the manipulation to fail. Tracking is therefore an important component in a robust grasping and manipulation framework. If the range sensing and segmentation methods described in Chapters 3 and 4 could be performed with sufficient speed, the tracking could be implemented by continuously repeating this process. Unfortunately, the current measurement rate (up to one minute per range scan) renders this approach unsuitable for real-time tracking. However, the textured polygonal models and initial pose information from range data segmentation present an ideal basis for 3D model-based tracking. To close the visual feedback loop, this chapter now addresses the problem of continuously updating the pose of modelled objects.

In Chapter 6, robust tracking of the end-effector is achieved by fusing kinematic and visual measurements and exploiting the fact that that gripper is a known target. Conversely, tracking unknown objects must rely solely on the detection of natural cues. Object tracking is further complicated by lighting variations, background clutter and occlusions that are likely to occur in domestic scenes. Furthermore, arbitrary objects may contain too few or too many visual features for robust matching. Conventional model-based tracking algorithms are typically based on a single class of visual cue, such as intensity edges or texture, and fail when visual conditions eventually become unsuitable (as will be demonstrated in Section 5.5). However, it is also observed that different cues often exhibit independent and complementary failure modes. The problem of unpredictable visual conditions can therefore be alleviated by fusing multiple visual cues. Multi-cue fusion achieves tracking robustness by reducing the likelihood that all visual features fail simultaneously, allowing the tracker handle a wide variety of viewing conditions. For example, if edge detection fails due to a low contrasting background, the tracking filter can simply rely on cues such as colour and texture.

The following section examines the motivation for multi-cue visual tracking in greater detail and reviews the common approaches to the problem. An overview of the proposed framework is presented in Section 5.2, in which objects are tracked by fusing colour, texture and edge cues in a Kalman filter. The filter is described in Section 5.3, and Section 5.4 follows with details of the image processing and measurement model for each cue. Finally, Section 5.5 provides implementation details and experimental results. The performance of the proposed multi-cue algorithm is compared with single-cue tracking filters to validate the increased robustness gained through fusion. Video clips of the experimental tracking results can be found in the *Multimedia Extensions*, along with source code to implement the algorithm.

5.1 Introduction

Visual tracking is employed in a variety of robotic applications that require the localization of dynamic targets, including mobile robot navigation [30], human-machine interaction [72] and machine learning [11]. Clearly, visual tracking is also necessary in domestic manipulation tasks involving moving objects. In practice, domestic objects are actually often static[1], and a well calibrated robot could employ a *sense-then-move* strategy (initial pose estimation followed by blind grasp) to perform many tasks, as demonstrated in [10]. In practice, *sense-then-move* without visual tracking is defeated by at least two sources of uncertainty: kinematic and camera calibration errors, and dynamics imposed by the robot on the environment, as described below.

It was suggested in Chapter 1 that reliance on accurate calibration in domestic robotics may be unsustainable or even undesirable, and emphasis should therefore be placed on robustness to calibration errors. In the case of visual localization, calibration errors introduce a varying bias in the estimated pose. As the active cameras move, any stationary object will appear to undergo an opposite motion on the image plane. When transformed from the camera frame to the world frame (see Figure 2.5), the estimated pose of the object should remain constant despite the camera motion. In practice, ego-motion creeps in as the transformation (equation (2.30)) is subject to kinematic calibration errors. Motion of the active head therefore reduces any confidence in the previously sensed location of an object, whether static or dynamic. Unstable grasps or collisions during the execution of a task can impose additional dynamics and further increase the uncertainty in the expected pose of target objects.

Visually tracking target objects reduces these uncertainties, and therefore plays an important role in hand-eye systems even in static domestic environments. As already noted, 3D model-based tracking is particularly suited to the system developed in this monograph since object modelling (see Chapter 4) provides textured polygonal models and accurate initial pose information.

A variety of approaches to object tracking have been proposed in the literature, and algorithms are typically based on the detection of a particular cue, such as colour [11], edges [89,98,152] or feature templates [113,114]. However, individual cues can only provide reliable tracking under limited conditions. For example,

[1] In fact, this is a requirement of light stripe scanning (see Chapter 3).

motion and colour require the presence of contrasting foreground and background colours, and only recover partial pose information. Intensity edges similarly require a highly contrasting background, and are distracted by surface textures [89]. View-based approaches such as template tracking are distracted by reflections and lighting variations, while some objects contain insufficient texture for reliable matching. Single-cue tracking is therefore unsuitable in an unpredictable domestic setting, since the above failure modes will inevitably be encountered at some time.

To address the problem of reliable tracking in unpredictable conditions, it should be observed that different cues are subject to independent failure modes. Thus, the simultaneous failure of all cues is much less likely than the loss of any single modality. Robustness can be improved by exploiting multi-cue visual tracking, which is the basis of the framework proposed in this chapter. A 3D model-based algorithm is developed to track textured polygonal object models in stereo video streams by fusing colour, edge and texture cues in a Kalman filter framework.

Multi-cue tracking schemes generally adopt one of two basic approaches: *sequential cue selection* or *cue integration*. In the sequential cue selection framework, the tracker may switch between a number of cues, but only one cue contributes to the estimated state at any time. In cue integration, all cues are tracked simultaneously and fused to estimate the state. Kragić and Christensen [87] present a classical two-stage sequential cue algorithm to estimate the pose of an object for position-based visual servoing. A view-based template matching algorithm is first applied to select a coarse region of pose space, after which the object is tracked using edge cues.

Toyama and Hager [153] generalize the sequential cue approach to a hierarchy of *selector* and *tracker* algorithms. The hierarchy orders algorithms in terms of precision, although some may operate on the same visual cue. Selectors are defined as algorithms that may produce a false positive (erroneously identify a feature as part of the object), while *trackers* never produce false positives. An associated state transition graph determines which algorithm is active depending on current tracking performance. For robotic applications, Prokopowicz *et al.* [125] suggest that knowledge of the expected behaviour of visual features for the given task and environment should also be incorporated into the cue selection algorithm. Darrell *et al.* [29] present an alternative approach to cue selection for tracking people in 2D images using parallel rather than sequential cue processing. Skin colour, range segmentation and face detection are processed independently for each frame, and state estimation is based on an order of preference between detected cues.

The main drawback of sequential cue selection in multi-cue tracking is that considerable visual information is simply discarded in any given frame. Furthermore, accurate evaluation of tracking performance (in particular, detection of false positives) is required to guide the selection of the most suitable cue for current tracking conditions. Conversely, the approach of cue integration avoids both of these issues, but at the expense of greater computational load. In a typical cue integration scheme, the pose of the object is estimated independently from each cue modality, and the results are combined using a suitable fusion operator. A variety of operators have been proposed in practice, and a review of common approaches is given in [13].

Spengler and Schiele [141] present two typical Bayesian decision approaches to cue integration: democratic voting and particle filtering. In both cases, each cue maps the input image into a conditional probability density for the state of the target. Democratic voting then calculates the combined state as the maxima of the weighted sum of the probability densities, where the weights are based on the past performance of cues. Particle filtering (also known as the CONDENSATION algorithm in visual tracking literature) fuses state probability densities in a similar manner, but represents the combined density as a small number of sampled states (known as particles). Particles are easily clustered to give multiple state hypotheses, which enables tracking of multiple targets and provides robustness to association errors.

Cue integration can also be performed without explicitly modelling conditional probability densities, and two such tracking schemes are presented in [87]: consensus voting and fuzzy logic. Consensus voting, similar to the Hough transform, operates by discretizing state space and allowing each cue to vote for zero or more states. The supporting cues for the most popular state are then combined to refine the estimated state. Fuzzy logic operates by assigning to each state a membership value corresponding to each cue. Fusion is performed by maximizing the combined membership values using a min-max fuzzy logic operator. Uhlin *et al.* [157] present an alternative non-Bayesian fusion framework for the specific task of tracking an arbitrary moving target represented by a binary image mask. Motion detection identifies pixels which are moving independently of the background, and disparity segmentation distinguishes pixels that are moving consistently with the current target. The new pixels are fused with the existing mask using simple binary logic operations.

The ICONDENSATION algorithm proposed by Isard and Blake [72] is a notable compromise between the cue selection and integration schemes. In this work, a human hand is tracked in a particle filtering framework by fusing skin colour and shape template features. Skin colour detection forms the basis of an importance function from which the particles are re-sampled at each update, while shape detection contributes to the actual estimated state. Thus, while skin colour is not used directly to estimate the state, both cues influence the final outcome.

The multi-cue integration techniques described above work well for feature-based applications in which the estimated state is a 2D position on the image plane, but are less suited to 3D model-based tracking. In particular, fusion operators such as voting and ICONDENSATION quickly become computationally intractable as the number of estimated parameters increase. Furthermore, both approaches typically require each cue to estimate the state independently of other cues. While this is usually possible in 2D tracking, cues such as motion and colour do not sufficiently constrain the 3D pose estimation problem.

The framework proposed in the chapter overcomes these issues by integrating multiple visual cues using the Kalman filter framework, which is commonly exploited in sensory fusion applications such as mobile robot navigation [84]. In this approach, each cue has an associated measurement model to predict the observations for a given state. In each measurement cycle, the estimated state is iteratively updated by minimizing the weighted error between the observed and predicted measurements. Thus, any cue with a suitable measurement model can contribute to the estimated state since *explicit* pose reconstruction is not necessary. Like other Bayesian approaches, the Kalman filter also maintains an estimate of the state covariance, which is useful for evaluating tracking performance. Furthermore, experimental results will demonstrate that the Kalman filter imposes minimal computational expense compared to other components of the tracking framework, such as image processing. The proposed method is now described in detail.

5.2 System Overview

The aim of visual tracking is to estimate the trajectory of an object in a sequence of stereo images, given an initial estimate of the pose. The cues used for multi-cue tracking are colour centroid, intensity edges and surface texture, which are chosen to provide complementary failure modes and redundant pose information. The pose of the object in the kth frame is represented by a 6 dimensional pose vector $\mathbf{p}(k) = (X, Y, Z, \phi, \theta, \psi)^\top$ (see Section 2.1.4). The tracked object is modelled as a 3D polygonal mesh with textured facets, as detailed in Chapter 4. It will be demonstrated below that a polygonal model provides a reasonable approximation for objects with either flat or curved surfaces. The N vertices of the facets are represented by the set of points $V = \{^O\mathbf{V}_i : i = 1, \ldots, N\}$.

Figure 5.1 shows a typical frame from the video sequence with a moving rectangular box, which will be used to illustrate the discussion below and in later sections. Tracking is divided into three sub-tasks: *feature prediction*, *detection/association* and *state update*. Feature prediction consists of estimating the expected visual cues from the textured 3D model and predicted pose. As usual, the pose is predicted by evolving the estimated state from the previous frame according to the dynamics model in the tracking filter. Using the central projection camera model introduced in Section 2.2.1, the projected vertices $^{L,R}\hat{\mathbf{v}}_i$ of the object in the predicted pose are calculated as:

$$^{L,R}\hat{\mathbf{v}}_i(\hat{\mathbf{p}}) = {}^{L,R}\mathrm{P}\,{}^W\mathrm{H}_O(\hat{\mathbf{p}})\,{}^O\mathbf{V}_i \tag{5.1}$$

where $^{L,R}\mathrm{P}$ are the camera projection matrices (given by equation (2.29)) and $^W\mathrm{H}_O(\hat{\mathbf{p}})$ is the transformation from the object frame to the world frame for the predicted pose $\hat{\mathbf{p}}$. For the test sequence in Figure 5.1, the predicted pose of the box is indicated by the wireframe overlay. To predict the visual cues, the 3D model of the target is rendered into an image buffer using standard computer graphics techniques. The synthetic images in Figures 5.2(a) and 5.2(b) show the expected appearance of textures and edges for the predicted pose of the tracked box.

The predicted visual features are used to guide the detection and association of real cues in the captured frame. A rectangular *region of interest* (ROI) for visual measurements is calculated as the bounding box containing all the projected vertices of the model (with a fixed margin around the points). The ROI for the test frame in Figure 5.1 is indicated by the white rectangle around the object. To reduce computational load and background distractions, the colour centroid, texture and edge cues are only calculated from pixels within the rectangular ROI.

Fig. 5.1. Captured image, predicted pose and region of interest (ROI)

(a) Predicted appearance of texture

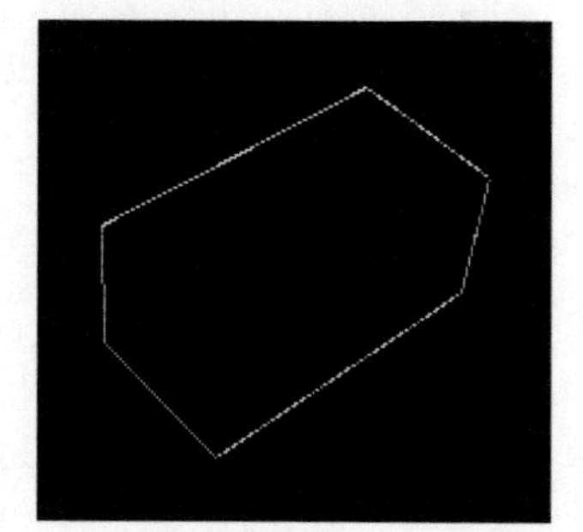

(b) Predicted appearance of edges

Fig. 5.2. Predicted appearance of object.

The prediction, detection and association algorithms for each cue are detailed in Section 5.4. In brief, the colour centroid is predicted as the mean position of projected vertices, and measured by processing the ROI with a colour filter constructed from the appearance of the object in the initial frame. To measure edges, the ROI is first processed with an edge detection filter, and the resulting edge pixels are matched to the predicted edges shown in Figure 5.2(b). The edges are parameterized by fitting a line to the pixels associated with each boundary segment. For texture cues, the synthetic image in Figure 5.2(a) is processed using a texture quality filter to locate salient features. The predicted features are then matched to the captured frame (using sum of squared difference minimization) to find the corresponding real texture cues. Finally, all the measurements are fused to update the estimated pose of the object.

Many tracking algorithms achieve continuity by associating new measurements with a fixed set of features over several frames. The association problem becomes more difficult as the appearance of features change with time, until eventually the features must be replaced with a new set. Conversely, model-based tracking allows for the selection of new features with *every* frame, since continuity is provided by the underlying object model. The temporal association problem is replaced with the simpler problem of associating predicted and measured features. Using each cue only once eliminates the need to explicitly update the appearance and assess the quality of features over time (as used by [138]). Furthermore, when feature detection and association errors occasionally arise, the offending features do not perturb the estimated pose for more than a single frame.

5.3 Kalman Filter Framework

The Kalman filter is a well known algorithm for optimally estimating the state of a linear dynamic system from noisy measurements, and is used for many tracking applications in both vision and robotics [84, 163]. The Kalman filter also provides a statistically robust framework for fusing different measurement modalities, which is exploited in the framework developed below. Furthermore, the filter maintains an estimate of the uncertainty in the tracked parameters, which can be useful for evaluating tracking performance. The *Iterated Extended Kalman Filter* (IEKF) extends the framework to non-linear systems, such as the present application, by solving the filter equations numerically. Appendix C describes the basic equations of the IEKF, and a detailed treatment of Kalman filter theory can be found in [8, 74].

The variables we wish to estimate are encapsulated as the *state* of the tracking filter. For 3D object tracking, the state is a 12 dimensional vector $\mathbf{x}(k) = (\mathbf{p}(k), \dot{\mathbf{r}}(k))^\top$, where $\mathbf{p}(k)$ and $\dot{\mathbf{r}}(k)$ represent the pose and velocity screw of the target in the kth video frame. The IEKF algorithm to estimate the state can be summarized in two broad steps: *state prediction* based on a known dynamic model, followed by *state update* based on the latest measurements. Assuming the object has smooth motion, the evolution of the state is modelled by constant velocity dynamics:

$$\mathbf{p}(k+1) = \mathbf{p}(k) + \dot{\mathbf{r}}(k)\Delta t \tag{5.2}$$

$$\dot{\mathbf{r}}(k+1) = \dot{\mathbf{r}}(k) \tag{5.3}$$

where Δt is the sample time between frames. The first step in the IEKF algorithm predicts the current state by applying this model to the previous state estimate.

In the update step, cues are fused with the predicted state through measurement models that predict the observed features as a function of the predicted state. The new state estimate is recovered by minimizing the weighted observation error between the predicted and measured features. Colour, edge and texture patches extracted from the captured stereo images are stored in a measurement vector $\mathbf{y}(k+1)$, which has variable dimension depending on the number of detected features. A measurement model predicts these measurements $\hat{\mathbf{y}}(k+1)$ given the predicted state $\hat{\mathbf{x}}(k+1|k)$ in the $(k+1)$th frame as:

$$\hat{\mathbf{y}}(k+1) = \mathbf{h}(\hat{\mathbf{x}}(k+1|k)) + \mathbf{w}(k+1) \tag{5.4}$$

where $\mathbf{w}(k+1)$ is an estimate of the measurement noise. The non-linear measurement equations, $\mathbf{h}(\mathbf{x})$, are detailed in Section 5.4 for each cue. For a linear Kalman filter, the current state $\hat{\mathbf{x}}(k+1)$ is then estimated as the sum of the predicted state and observation error (difference between the predicted and observed measurement vectors), weighted by the Kalman gain. The IEKF differs from the linear Kalman Filter in that the updated state is computed numerically via iterative re-linearization of the non-linear measurement model. Details of the IEKF equations can be found in Appendix C.

The orientation component of the state vector must be handled carefully in the IEKF as Euler angles are non-unique and even degenerate for some orientations (see Section 2.1.3). To address this issue, the Euler angles in the state vector represent only the inter-frame differential orientation, while the total orientation is stored as a quaternion outside the IEKF [15, 161]. At each state update, the IEKF estimates the differential orientation between the new pose and the external quaternion. The differential Euler angles are then integrated into the external quaternion and reset to zero for the next measurement cycle.

As noted earlier, the IEKF framework is extensible; any modality with a suitable measurement model can be integrated, including cues that provide incomplete pose information. The same mechanism allows features to be fused from multiple cameras by supplying a measurement model for each image plane. Unlike conventional stereo reconstruction, this approach eliminates the need to search for corresponding features as all measurements are coupled implicitly through the filter state. Camera failures are also tolerated as tracking continues even if the object is obscured from all but one camera.

5.4 Feature Measurement

As described earlier, the robustness of multi-cue tracking arises from integrating cues with independent and complementary failure modes. For this purpose, the proposed filter is based on colour centroid, intensity edges and texture templates. As described in Section 5.2, the measurement process begins by predicting the pose of the object in the current frame and identifying the ROI. Each modality then requires a detection process to extract features from the ROI (in both stereo fields) before assembling the results into a measurement vector. Finally, the IEKF requires a measurement model associated with each modality to predict the observed features for a given pose. The following sections describe the detection process and measurement model for each cue. For pedagogical purposes, the frame shown in Figure 5.1 will be used to illustrate the detection algorithms.

5.4.1 Colour

The colour cue measures the centroid of pixels in the captured frame that are congruent with the expected colour of the object. The measurement process is implemented

(a) ROI in initial image with outline showing pixels used for creation of object colour filter

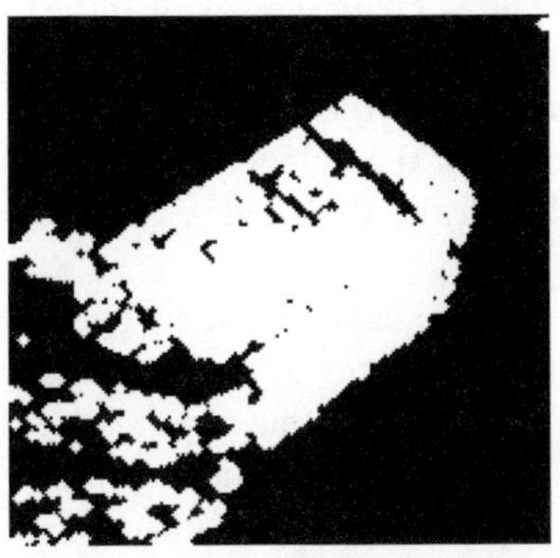

(b) Output of colour filter applied to current frame ROI

(c) Measured colour centroid (black square)

Fig. 5.3. Colour cue initialization and measurement of colour centroid.

using a simple colour filter constructed from an image of the object, which is sufficient for tracking in scenes with minimal lighting variations. Adaptive colour tracking techniques [2, 101] provide greater robustness to lighting variations and clutter and could easily replace the colour filter described below.

One of the difficulties associated with colour filtering is the possibility that the background contains distracting colours similar to those in the target. Thus, the filter initialization algorithm is designed to promote unique colours in the target while suppressing those colours common to the target and background. This is achieved by exploiting the accuracy of the initial pose estimate furnished by range data segmentation. The filter is constructed from the ROI in the initial frame of the tracking sequence, and then applied to every subsequent frame.

For our box tracking example, the initial frame in the sequence and associated ROI are shown in Figure 5.3(a). The boundary of the object is identified by projecting the initial pose of the model onto the image plane (equation (5.1)), as shown by the white outline in Figure 5.3(a). Colour information is compiled in an RGB histogram by partitioning RGB space into uniform cells (16 cells per channel was found to be sufficient). The histogram is compiled from *all* pixels in the ROI both inside and outside the boundary of the object; pixels inside the object contribute positively to the histogram while background pixels contribute negatively. Let H_i represent the set of colours associated with the ith cell in the histogram, and h_i represent the associated tally. Furthermore, let $C(x,y)$ represent the colour of the pixel located at (x,y) in the captured frame, and O represent the set of pixel locations within the object boundary. Then, the accumulated count for the ith histogram bin is

$$h_i = \sum_{(x,y)\in ROI} \begin{cases} +1 \text{ if } (x,y) \in O,\ C(x,y) \in H_i \\ -1 \text{ if } (x,y) \notin O,\ C(x,y) \in H_i \\ 0 \ \text{ otherwise} \end{cases} \tag{5.5}$$

After accumulating all pixels, cells with a negative count are set to zero. Outlier rejection is applied by iteratively eliminating the cell with the lowest count until no less than 90% of the original total tally remains. Finally, the colours associated with the remaining non-zero cells form the pass-band of the colour filter. The filter is stored as a binary look-up table indicating whether the RGB index value occupies the pass-band.

During tracking, the colour filter is applied to the captured frame by replacing each pixel in the ROI with value in the look-up table. Figure 5.3(b) shows the result of colour filtering for the ROI in Figure 5.1. The centroid of the object is recovered by applying binary connectivity to the output of the filter and calculating the centroid of the largest blob. Finally, the IEKF requires a measurement function to predict the colour centroid for a given state. The colour centroid is approximated as the projected centroid of the 3D model, after transforming to the given state:

$$^{L,R}\hat{\mathbf{c}}(\mathbf{x}) = {}^{L,R}\mathrm{P}^{W}\mathrm{H}_O(\mathbf{x}) \sum_{\mathbf{V}_i \in V} \mathbf{V}_i \tag{5.6}$$

where $^{L,R}\mathrm{P}$ are the camera projection matrices, $^{W}\mathrm{H}_O(\mathbf{x})$ is the transformation from the object frame to the world frame for state $\mathbf{x}$, and $^{L,R}\hat{\mathbf{c}}(\mathbf{x})$ are the predicted centroids on the left and right image planes. This prediction equation actually introduces a systematic bias due to the non-linearity of the projective transformation; the projection of the mean point (ie. the prediction) is not equal to the mean of the projected points (ie. the image plane measurement). However, this effect is neglected since it is much smaller than the random error arising from noise in the output of the colour filter, and in any case the bias is readily compensated by the other visual cues.

5.4.2 Edges

The predicted edges for the tracking scenario in Figure 5.1 are shown in Figure 5.2(b). Only the jump boundaries outlining the object are considered in the detection process, since internal edges are easily distracted by surface textures (see Figure 5.4(a)). The jth predicted edge segment is represented by the 3D vertices $\mathbf{A}_j \in V$ and $\mathbf{B}_j \in V$ of the end-points and their associated projections $\mathbf{a}_j(\mathbf{x})$ and $\mathbf{b}_j(\mathbf{x})$ on the image plane for the current pose (via equation (5.1)). For each pair of end-points, the normal direction $\mathbf{n}_j = (n_{xj}, n_{yj})^\top$, orientation θ_j and perpendicular distance d_j to the image plane origin are calculated as:

$$\mathbf{n}_j = \begin{pmatrix} 0 & -1 \\ 1 & 0 \end{pmatrix} (\mathbf{b}_j - \mathbf{a}_j) \tag{5.7}$$

$$\theta_j = \tan^{-1}(n_{yj}/n_{xj}) \tag{5.8}$$

$$d_j = -\mathbf{n}_j^\top \mathbf{a}_j \tag{5.9}$$

These parameters guide the detection of intensity edges in the captured frame. As usual, the first step in edge detection is to calculate directional intensity gradients, $g_x(x,y)$ and $g_y(x,y)$. A central difference approximation to these gradients is

$$g_x(x,y) = I(x+1,y) - I(x-1,y) \tag{5.10}$$

$$g_y(x,y) = I(x,y+1) - I(x,y-1) \tag{5.11}$$

where $I(x,y)$ is the intensity channel. An edge is detected at pixel position $\mathbf{e}_k$ when $g_x(\mathbf{e}_k) > e_{th}$ and $g_y(\mathbf{e}_k) > e_{th}$ for the gradient threshold e_{th}, and the orientation of the detected edge is calculated as $\theta_k = \tan^{-1}(g_x/g_y)$.

Figure 5.4(a) shows the output of the edge detector applied to the ROI in Figure 5.1. Clearly, raw edge detection produces numerous spurious measurements due to variations in lighting and surface texture. These distractions are eliminated by combining edge detection with the output of the colour filter (see Figure 5.3). First, the colour filter output is scanned to identify the pixels on the convex boundary of the largest connected region. Edge pixels beyond a threshold distance of these boundary pixels are eliminated from the output of the edge detector. The remaining edge pixels are treated as possible candidates for the desired jump boundary outlining the object. Finally, skeletonization is applied to reduce the edges to the minimal set of candidates, shown in Figure 5.4(b).

Following image processing, an association algorithm attempts to match each edge pixel to one of the predicted line segments. The kth edge pixel, at position $\mathbf{e}_k$ and with orientation θ_k, is associated the the jth line segment when the following conditions are satisfied:

$$\mathbf{n}_j^\top \mathbf{e}_k + d_j < r_{th} \tag{5.12}$$

$$|\theta_j - \theta_k| < \theta_{th} \tag{5.13}$$

$$0 \le (\mathbf{e}_k - \mathbf{a}_j)^\top (\mathbf{b}_j - \mathbf{a}_j) \le |\mathbf{b}_j - \mathbf{a}_j|^2 \tag{5.14}$$

Condition (5.12) requires the candidate pixel to satisfy the line equation to within an error threshold r_{th}. Condition (5.13) enforces a maximum angular difference θ_{th} between the orientation of the edge pixel and predicted line segment. Finally, condition (5.14) ensures that the pixel lies within the end points of the segment. When the pixel matches more than one predicted edges, the ambiguity is resolved by assigning the candidate to the segment that gives the minimum residual error in condition (5.12).

After the association step, the jth predicted segment has m_j matched pixels. For segments with sufficient matched points ($m_j > m_{th}$), two measurements are generated

(a) Edge filter output (grey-level indicates edge orientation)

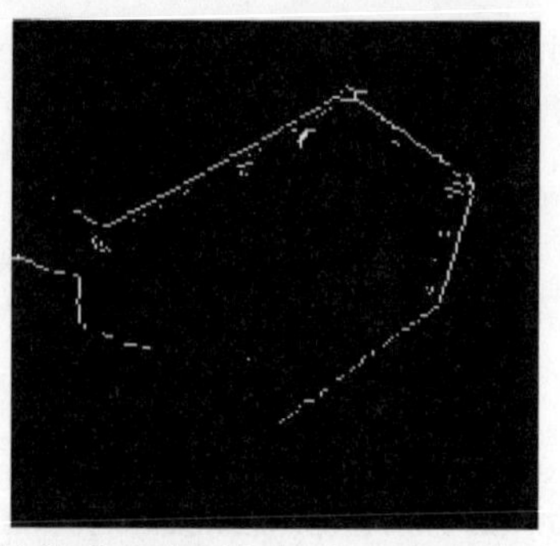

(b) Skeletonized jump boundary pixels

(c) Detected edges (line indicates orientation and square indicates mean position)

Fig. 5.4. Edge detection and matching.

for the IEKF: the average position of edge pixels $\mathbf{m}_j$, and the normal orientation θ_j of a fitted line using principal components analysis (PCA) [51]. The robustness to mis-matched points is improved by applying an initial PCA, rejecting points with a residue greater than two standard deviations, and re-applying PCA to the remaining points. Figure 5.4(c) shows the mean position and orientation of detected edges for the tracking example in Figure 5.1.

The mean of pixels associated with the jth segment is unlikely to coincide with the midpoint between projected end-points $\hat{\mathbf{a}}_j$ and $\hat{\mathbf{b}}_j$, as a result of occlusions and other distractions. Furthermore, the observed mean is difficult to predict for a given state, so $\mathbf{m}_j$ is not a useful measurement for the IEKF. Instead, the distance r_j between $\mathbf{m}_j$ and the measured edge, which is identically zero and invariant to occlusions, is added to the measurement vector. The mean position $\mathbf{m}_j$ is supplied as an input to the IEKF, and the measurement model predicting $\hat{r}_j(\mathbf{x})$ for a given state $\mathbf{x}$ is:

$$\hat{r}_j(\mathbf{x}) = \hat{\mathbf{n}}_j^\top(\mathbf{x})[\mathbf{m}_j - \hat{\mathbf{a}}_j(\mathbf{x})] \tag{5.15}$$

where the normal vector $\hat{\mathbf{n}}_j(\mathbf{x})$ of the segment is calculated from equation (5.7). Since the measured r_j is identically zero, the pose estimated by the IEKF minimizes the perpendicular distance between the measured mean $\mathbf{m}_j$ and the predicted edge.

To predict the orientation θ_j, the IEKF is supplied with the end-points $\mathbf{A}_j$ and $\mathbf{B}_j$ of the associated segment in the 3D model. The image plane projections $\hat{\mathbf{a}}_j(\mathbf{x}) = (\hat{x}_{aj}, \hat{y}_{aj})^\top$ and $\hat{\mathbf{b}}_j(\mathbf{x}) = (\hat{x}_{bj}, \hat{y}_{bj})^\top$ for a given state $\mathbf{x}$ are then calculated using equation (5.1), the orientation is predicted as:

$$\hat{\theta}(\mathbf{x}) = \tan^{-1}\left(\frac{\hat{x}_{bj} - \hat{x}_{aj}}{\hat{y}_{aj} - \hat{y}_{bj}}\right) \tag{5.16}$$

In summary, edge cues are parameterized by θ_j and $r_j \equiv 0$, and equations (5.15) and (5.16) serve as the measurement model for the IEKF, with inputs $\mathbf{A}_j$, $\mathbf{B}_j$ and $\mathbf{m}_j$. It should be noted that the measurement vector could have been parameterized using the distance to the origin instead of the distance to the measured mean point. However, the error variance of the distance to the origin varies significantly with the position and orientation of the line segment. Conversely, the measurement error variance for the distance to the mean point is small and independent of the position and orientation of the segment. The latter formulation therefore leads to a simpler calculation of the measurement error covariance matrix required by the tracking filter.

5.4.3 Texture

Texture tracking algorithms typically represent cues as small greyscale image templates, and match these to the captured image using sum of squared difference (SSD) or correlation-based measures. The templates are generally view-based and therefore dependent on the pose of the object. Most tracking algorithms address the issue of view-dependence by maintaining a quality measure to determine when a template no longer represents the current appearance of a feature. Conversely, the use of a textured 3D object model allows a new set of view-dependent features to be predicted in every frame. While this approach is computationally expensive, the feature selection calculations can be optimized to operate in real-time. As noted earlier, model-based 3D tracking also removes the need for stereo reconstruction from corresponding measurements. As with the other cues, the measurement process described below is applied independently to each camera.

Figure 5.2(a) shows the predicted appearance of textures for the object in Figure 5.1. Clearly, some regions constrain the tracking problem better than others; areas exhibiting omni-directional spatial frequencies such as corner and salt-and-pepper textures are generally considered the most suitable. A widely accepted technique for locating such features is the quality measure proposed by Shi and Tomasi [138]. For each pixel in the rendered image, a matrix Z is computed as:

$$\mathrm{Z} = \sum_{\mathbf{x}\in W} \begin{pmatrix} g_x^2 & g_x g_y \\ g_x g_y & g_y^2 \end{pmatrix} \tag{5.17}$$

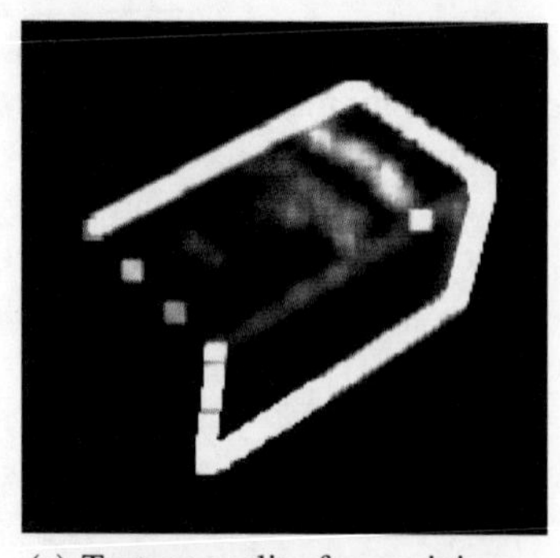

(a) Texture quality from minimum eigenvalue of Z matrix.

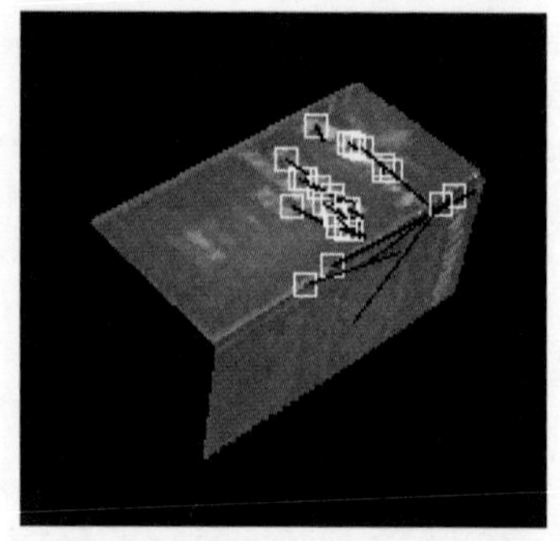

(b) Candidate templates (squares) and displacement vectors (lines, *x10* scale).

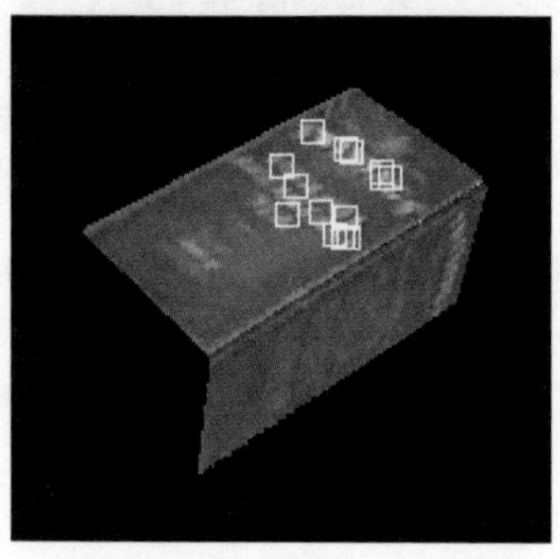

(c) Validated candidate templates

(d) Matched features in captured frame

Fig. 5.5. Texture feature selection and matching.

where g_x and g_y are spatial intensity gradients given by equations (5.10)-(5.11), and W is the $n \times n$ template window centred on the pixel. Good texture features are identified as satisfying

$$\min(\lambda_1, \lambda_2) > \lambda_{th} \tag{5.18}$$

where λ_1 and λ_2 are the eigenvalues of Z. Figure 5.5(a) shows the minimum eigenvalue at each pixel of the rendered object (Figure 5.2(a)). While the object is outlined by a high response to the quality measure, these areas usually straddle a jump boundary and are not suitable for tracking. Thus, the feature selector only examines pixels further than a window width from the jump boundary. A window-based search locates local maxima in the quality measure, which are assembled into a set of candidate cue locations. The ith candidate is associated with a position $\mathbf{m}_i$ and a window of pixels in the rendered image W_i (centred on $\mathbf{m}_i$) that define an intensity template.

Candidate templates are matched to the captured image using SSD minimization, which yields an offset $\mathbf{d}_i$ between the predicted and measured location of the ith cue. The SSD $\varepsilon_i(\mathbf{d})$ between the window centred at $\mathbf{m}_i$ in the rendered image and the window at $\mathbf{m}_i + \mathbf{d}$ in the captured image is evaluated as:

$$\varepsilon_i(\mathbf{d}) = \sum_{\mathbf{x}\in W_i} [J(\mathbf{x}) - I(\mathbf{x}+\mathbf{d}) - M(\mathbf{d})]^2 \tag{5.19}$$

where J is the rendered image, I is the captured image and $M(\mathbf{d})$ compensates for the mean difference between predicted and measured templates due to variations in lighting. This last term is given by:

$$M(\mathbf{d}) = \frac{1}{n^2} \sum_{\mathbf{x}\in W_i} [J(\mathbf{x}) - I(\mathbf{x}+\mathbf{d})] \tag{5.20}$$

The minimum SSD within a fixed search window D_i around the ith template determines the displacement $\mathbf{d}_i$:

$$\mathbf{d}_i = \mathrm{argmin}_{\mathbf{d}\in D_i} \varepsilon(\mathbf{d}) \tag{5.21}$$

Figure 5.5(b) shows the candidate features and measured displacement vectors (at $\times 10$ scale) for the object tracked in Figure 5.1. It can be expected that some candidate templates will be incorrectly matched to the captured image, since the rendered image is only an approximation. A two-stage validation process is therefore applied before adding texture cues to the measurement vector of the IEKF.

Let the initial candidate displacement vectors be represented by the set $C = \{\mathbf{d}_i\}$. Assuming the error between the predicted and actual pose of the object is approximately translational (a common requirement for template-based tracking), all correctly matched features will exhibit the same displacement. Thus, the first validation step finds the largest subset $C' \subseteq C$ containing approximately equal displacements. For each candidate displacement $\mathbf{d}_i$, a set of similar candidates S_i is constructed:

$$S_i = \{\mathbf{d}_j : ||\mathbf{d}_j - \mathbf{d}_i|| < d_{th}; \mathbf{d}_i, \mathbf{d}_j \in C\} \tag{5.22}$$

where d_{th} determines the maximum allowed distance between elements of S_i. After determining S_i for all candidates, a new set M_i is calculated for each candidate as the intersection of all sets of similar candidates that include $\mathbf{d}_i$:

$$M_i = \bigcap \{S_j : \mathbf{d}_i \in S_j\} \tag{5.23}$$

The new set M_i will contain only mutually supporting vectors such that $\mathbf{d}_i \in S_j$ and $\mathbf{d}_j \in S_i$ for all $\mathbf{d}_i, \mathbf{d}_j \in M_i$. Finally, the largest mutually supporting set of displacements $C' = \mathrm{argmax}|M_i|$ are taken as valid candidates.

A second validation test requires template matching to be invertible; for each vector $\mathbf{d}_i \in C'$, a template is extracted from the captured frame and matched to the rendered image by SSD minimization, giving a reverse displacement $\mathbf{r}_i$. Valid features should produce approximately equal forward and reverse displacements. Thus, the valid displacements are the subset $M \subseteq C'$ given by

$$M = \{\mathbf{d}_i : ||\mathbf{d}_i + \mathbf{r}_i|| < r_{th}, \mathbf{d}_i \in C'\} \tag{5.24}$$

where r_{th} is an empirically determined error threshold. Finally, the corresponding matched positions $\mathbf{m}_i + \mathbf{d}_i$ in the captured frame are added to the measurement vector

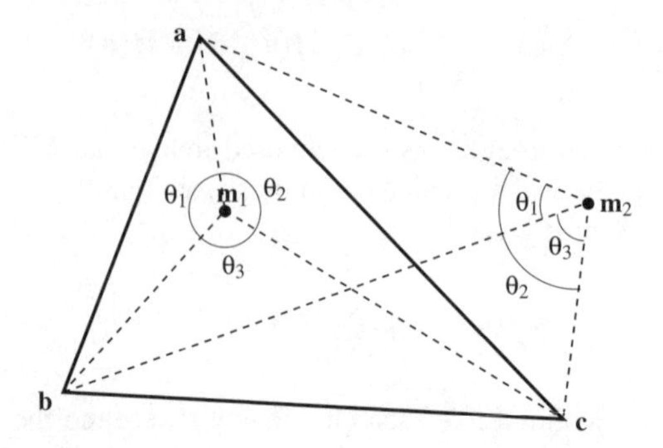

Fig. 5.6. Construction of the occupancy condition (5.25) for a facet given by vertices **a**, **b** and **c**. For point $\mathbf{m}_1$ within the facet, angles θ_1, θ_2 and θ_3 given by equations (5.26)-(5.28) sum to 2π. For point $\mathbf{m}_2$ outside the facet, the sum is less than 2π.

in the IEKF. Figure 5.5(c) shows the final set of validated features pruned from the initial candidates in Figure 5.5(b), and Figure 5.5(d) shows the location of matched features in the captured frame.

A measurement model to predict the location of texture features for a given state requires the construction of a model of 3D points $\mathbf{M}_i$ on the surface of the object corresponding to the projections $\mathbf{m}_i$ in the rendered image. Let $\mathbf{A}_j \in V$, $\mathbf{B}_j \in V$ and $\mathbf{C}_j \in V$ represent the 3D vertices of the jth triangular facet in the object model after transforming to a given pose, and $\mathbf{a}_j$, $\mathbf{b}_j$ and $\mathbf{c}_j$ represent their projections onto the image plane. Before calculating the 3D location of a texture feature, the facet containing the projection $\mathbf{m}_i$ must be identified. The following condition determines that $\mathbf{m}_i$ is bounded by the jth facet:

$$\theta_1 + \theta_2 + \theta_3 = 2\pi \tag{5.25}$$

where

$$\theta_1 = \cos^{-1}\left(\frac{(\mathbf{a}_j - \mathbf{m}_i)\cdot(\mathbf{b}_j - \mathbf{m}_i)}{|\mathbf{a}_j - \mathbf{m}_i||\mathbf{b}_j - \mathbf{m}_i|}\right) \tag{5.26}$$

$$\theta_2 = \cos^{-1}\left(\frac{(\mathbf{a}_j - \mathbf{m}_i)\cdot(\mathbf{c}_j - \mathbf{m}_i)}{|\mathbf{a}_j - \mathbf{m}_i||\mathbf{c}_j - \mathbf{m}_i|}\right) \tag{5.27}$$

$$\theta_3 = \cos^{-1}\left(\frac{(\mathbf{b}_j - \mathbf{m}_i)\cdot(\mathbf{c}_j - \mathbf{m}_i)}{|\mathbf{b}_j - \mathbf{m}_i||\mathbf{c}_j - \mathbf{m}_i|}\right) \tag{5.28}$$

Figure 5.6 illustrates the geometrical interpretation of the angles θ_1, θ_2 and θ_3 in condition (5.25). A search is performed over all visible facets to identify the bounding facet for each texture feature. The associated 3D vertices $\mathbf{A}_j$, $\mathbf{B}_j$ and $\mathbf{C}_j$ are then used to calculate the plane parameters of each bounding facet:

$$\mathbf{n} = (\mathbf{A}_j - \mathbf{B}_j) \times (\mathbf{C}_j - \mathbf{B}_j) \tag{5.29}$$

$$d = -\mathbf{n}^\top \mathbf{A}_j \tag{5.30}$$

where $\mathbf{n}$ is the normal direction and d is the distance to the origin of the camera frame. Finally, the 3D point $\mathbf{M}_i$ associated with the ith texture feature is given by the simultaneous solution of:

$$\mathrm{P}\mathbf{M}_i = \mathbf{m}_i \tag{5.31}$$

$$\mathbf{n}^\top \mathbf{M}_i + d = 0 \tag{5.32}$$

where equation (5.31) constrains $\mathbf{M}_i$ to the back-projected ray through $\mathbf{m}_i$ given the camera projection matrix P (equation (2.26)), and equation (5.32) constrains $\mathbf{M}_i$ to the surface of the facet. The solution $\mathbf{M}_i = (X_i, Y_i, Z_i)^\top$ in non-homogeneous coordinates is (with $\mathbf{m}_i = (x_i, y_i)^\top$ and $\mathbf{n} = (n_x, n_y, n_z)^\top$):

$$Z_i = -df/(n_x x_i + n_y y_i + n_z f) \tag{5.33}$$

$$X_i = x_i Z_i / f \tag{5.34}$$

$$Y_i = y_i Z_i / f \tag{5.35}$$

Finally, $\mathbf{M}_i$ is transformed from the left or right camera frame (depending on the camera in which $\mathbf{m}_i$ was measured) to the object frame:

$${}^O\mathbf{M}_i = {}^O\mathrm{H}_W(\mathbf{x})\,{}^W\mathrm{H}_{L,R}\,{}^{L,R}\mathbf{M}_i \tag{5.36}$$

where ${}^W\mathrm{H}_{L,R}$ is the transformation from the camera frame to the world frame, and ${}^O\mathrm{H}_W(\mathbf{x})$ is the transformation to the object frame for the given state $\mathbf{x}$. The surface point ${}^O\mathbf{M}_i$ is supplied as an input to the IEKF for each texture feature in the measurement vector. Finally, the predicted measurement $\widehat{\mathbf{m}}_i(\mathbf{x})$ for a given state $\mathbf{x}$ is simply calculated as the forward projection:

$$\widehat{\mathbf{m}}_i(\mathbf{x}) = {}^{L,R}\mathrm{P}\,{}^W\mathrm{H}_O(\mathbf{x})\,{}^O\mathbf{M}_i \tag{5.37}$$

where either ${}^L\mathrm{P}$ or ${}^R\mathrm{P}$ is applied depending on the camera in which the feature was measured (recalling that the cameras are processed independently). Since stereo correspondences are not required, the above framework is extensible from one to any number of cameras.

5.5 Implementation and Experimental Results

In this section, the performance of the proposed multi-cue tracking framework is examined in a number of real tracking scenarios. As described in Chapter 7, stereo images are captured at half PAL resolution (384×288 pixels) and PAL frame-rate, and projective rectification and radial distortion correction are applied to each frame prior to feature extraction. Image processing and Kalman filter calculations are performed on a dual 2.2 GHz Intel Xeon, and a number of optimizations are employed including parallel processing of stereo images in separate threads and MMX/SSE code optimizations. The current implementation achieves a processing rate of about 14 frames per second, and approximate processing times for each component of the

Table 5.1. Typical processing times for components of the multi-cue tracking algorithm.

Component	**Time** (ms)
Image capture and preprocessing	20
Colour measurement	5
Texture measurement	35
Edge measurement	7
IEKF update	5

algorithm are shown in Table 5.1 (values indicate total time for stereo frames). Texture features consume the greatest proportion of processing time due to the computationally expensive algorithms for perspectively correct rendering, texture quality measurement and normalized SSD minimization. Conversely, the IEKF fusion process imposes minimal computational load.

For the first three tracking scenarios, the performance of the multi-cue tracker is compared to the result achieved using single-cue trackers with edge only and texture only measurements[2]. The edge only and texture only trackers were implemented by simply discarding the unused features from the measurement vector in the IEKF. To enable a meaningful comparison, the video streams were captured from the experimental hardware and processed off-line for each tracking filter. Off-line processing also provides an indication of the tracking performance that would be achieved with a full frame-rate (25 Hz) implementation. Four tracking scenarios were examined: tracking in poor visual conditions (contrast and lighting), tracking in the presence of occlusions, tracking motion about an axis of symmetry, and tracking to overcome calibration errors in ego-motion. The results for each sequence are described in turn.

5.5.1 Sequence 1: Poor Visual Conditions

In the first tracking sequence, visual conditions were deliberately contrived to promote the loss of visual cues by introducing motion, low contrast, and lighting variations. The target object is a textured yellow box, and selected frames (right camera only) from the tracking sequence are shown in Figure 5.7 (see *Multimedia Extensions* for the full sequence). The box is initially placed against a highly contrasting background to enable reliable detection of all cues. The intensity edges are then weakened by moving the box to a low contrasting background. Finally, the box is moved back to its initial position and rotated to place the front face in shadow and impede the detection of texture cues. It should also be noted that the texture on the top face was not detected in this final orientation since it was hidden when the model was created. As described earlier, the sequence is processed using both single-cue and multi-cue tracking. Figure 5.7 shows the features observed by the multi-cue tracker, with the estimated pose of the box is overlaid as a wireframe model.

[2]Colour only tracking was not implemented since colour does not provide complete pose information.

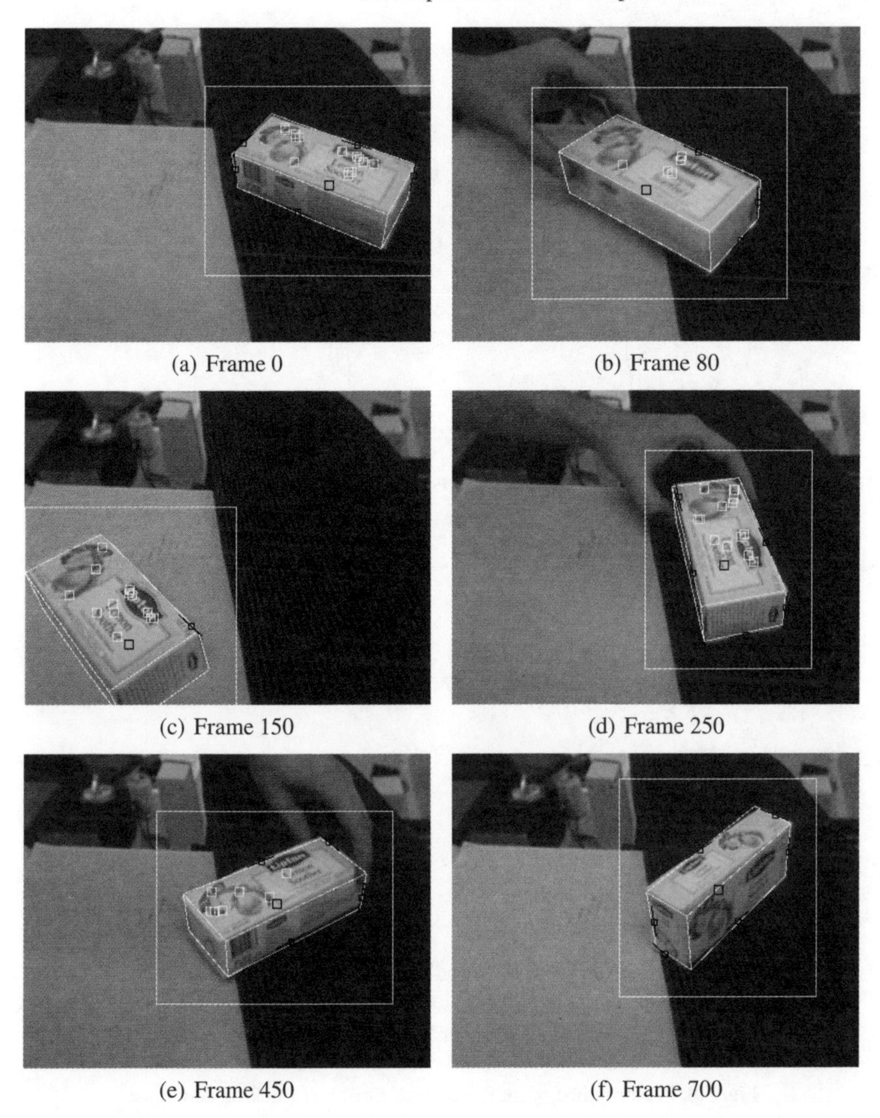

(a) Frame 0 (b) Frame 80

(c) Frame 150 (d) Frame 250

(e) Frame 450 (f) Frame 700

Fig. 5.7. Selected frames (right camera only) from box tracking sequence. Features used by the multi-cue tracker are overlaid along with the estimated pose. Frame numbers are only approximate (see *Multimedia Extensions* for the full sequence).

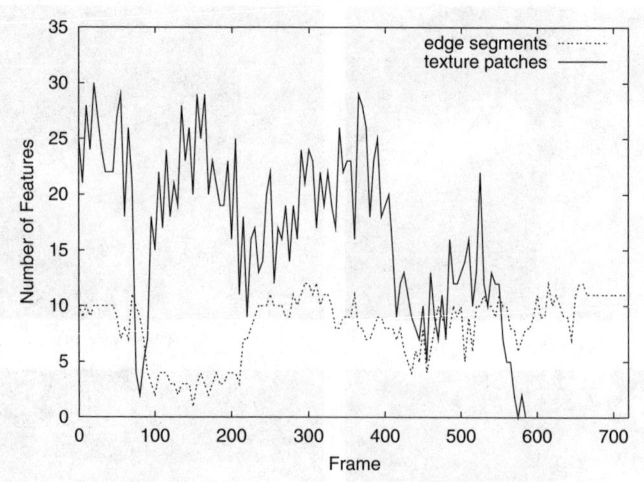

(a) Number of detected edge and texture cues (every fifth frame shown)

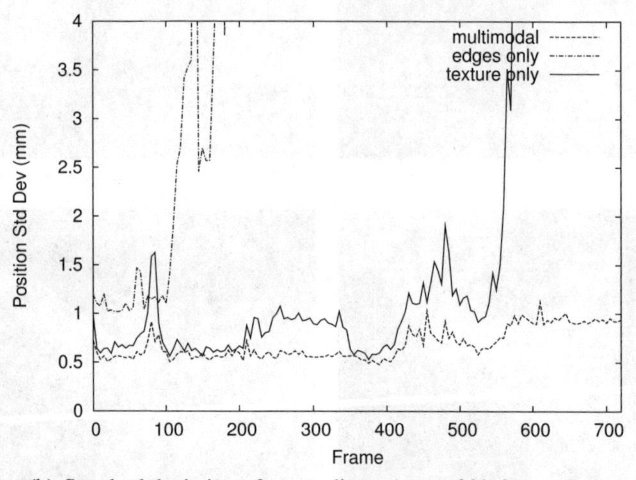

(b) Standard deviation of x-coordinate (every fifth frame shown)

Fig. 5.8. Performance of multi-cue tracker in box tracking sequence.

The plot in Figure 5.8 summarizes the number of edge and texture features observed for each frame in the sequence. As expected, the number of detected edges drops significantly between frames 100 and 200 when the box is placed in front of the low contrasting background. A sharp drop in the number of texture features is also evident between frames 70 and 100, which can be attributed to a blurring of the box as it is moved quickly to its new position (see Figure 5.7(b)). After frame 550, the number of texture features drops to zero as the altered lighting conditions hinder template matching. Despite the loss of individual cues at various times in the

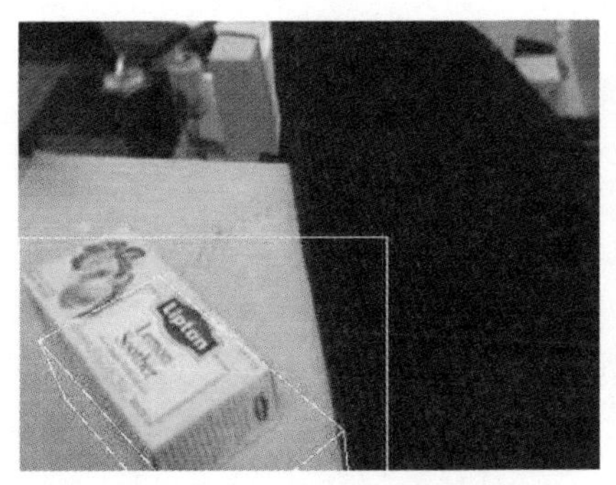

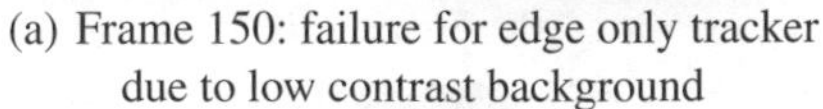

(a) Frame 150: failure for edge only tracker due to low contrast background

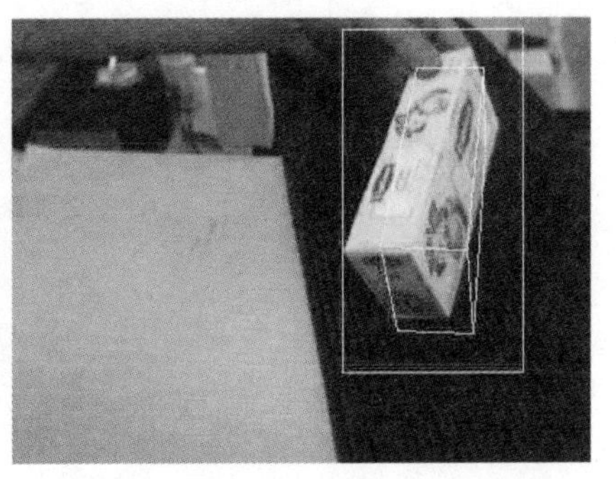

(b) Frame 560: failure for texture only tracker due to lighting variations

Fig. 5.9. Failure of single-cue tracking filters.

sequence, the multi-cue algorithm reliably tracks the box through the entire sequence by relying on available modalities.

To compare the performance multi-cue, edge only and texture only trackers, Figure 5.8(b) plots the standard deviation of the x-coordinate of the estimated pose for each tracker. As expected, multi-cue tracking yields a smaller variance in the estimated state than the single-cue trackers since it exploits all available constraints. The plot also highlights the failure of edge only and texture only tracking exhibited by divergence of the state covariance. As expected, failure of the edge tracker coincides with the loss of edge features due to the low contrasting background, and failure of the texture tracker occurs after the front face of the box is placed in shadow. The output of the edge and texture trackers at the point of failure is shown in Figure 5.9. Only the multi-cue tracker provides sufficient robustness in the presence of poor visual conditions to maintain track of the box for the entire sequence.

5.5.2 Sequence 2: Occlusions

The second tracking sequence also involves the loss of visual features, in this case caused by obstacles occluding the target. Selected frames from the tracking sequence (left camera only) are shown in Figure 5.10 (see *Multimedia Extensions* for the full sequence). Since the occlusions are introduced while the target is stationary, the initial 150 frames show the object in motion to demonstrate that the tracking filter is indeed estimating the pose and velocity without constraining the motion model. The first obstacle is introduced between frames 200 and 300 and occludes the texture cues on the upper surface of the box. Other obstacles are successively introduced between frames 400 and 700 to occlude most of the edges. Finally, the obstacles are removed and the object is again manipulated to verify the operation of the tracking filter.

Figure 5.11(a) plots the number of detected features in each frame for the multi-cue tracker. As expected, the first occlusion between frames 200 and 300 reduces the number of detected texture cues to zero (see Figure 5.10(c)). Similarly, occlusion

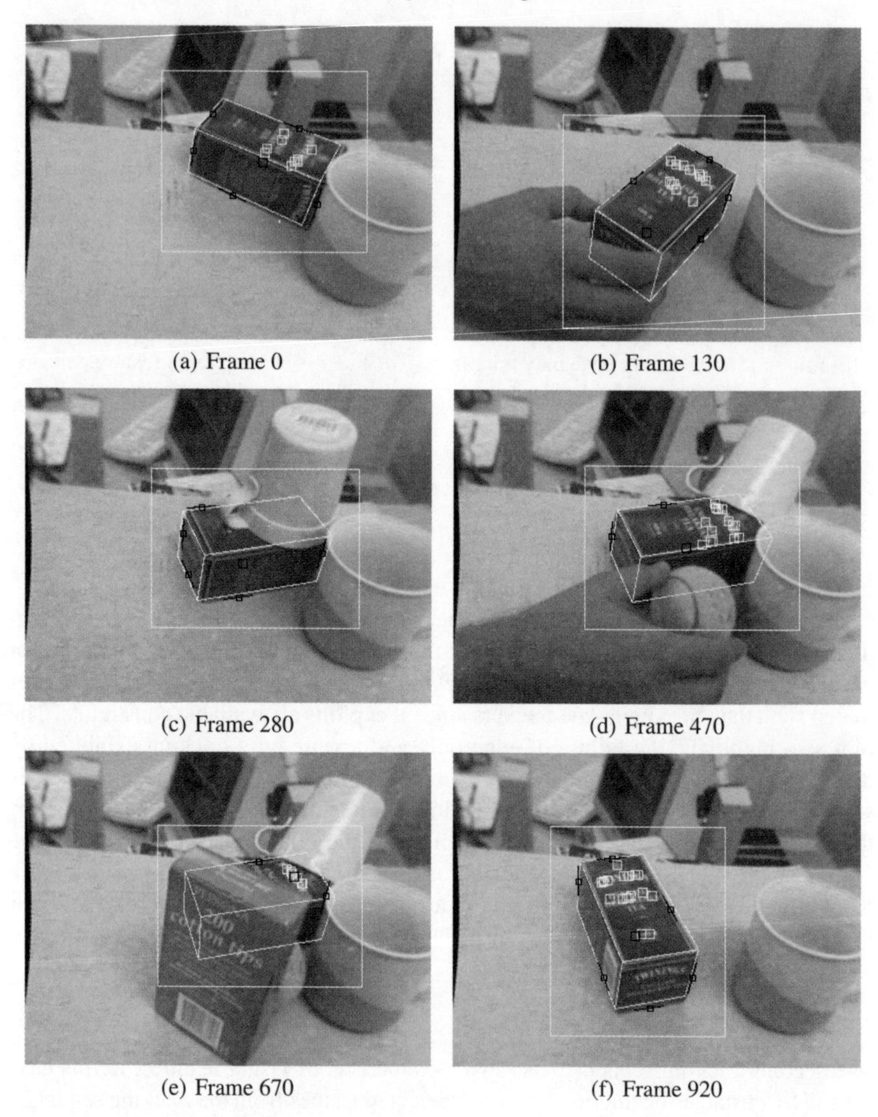

(a) Frame 0 (b) Frame 130

(c) Frame 280 (d) Frame 470

(e) Frame 670 (f) Frame 920

Fig. 5.10. Selected frames (left camera only) from occluded box tracking sequence. Features used by the multi-cue tracker are overlaid along with the estimated pose. Frame numbers are only approximate (see *Multimedia Extensions* for the full sequence).

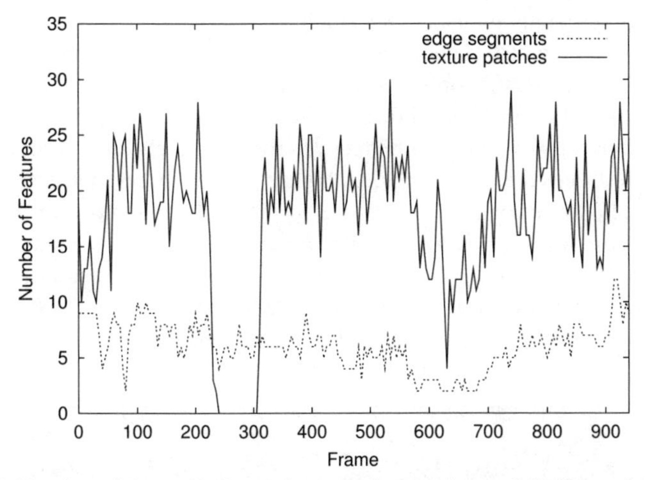

(a) Number of detected edge and texture features (every fifth frame shown)

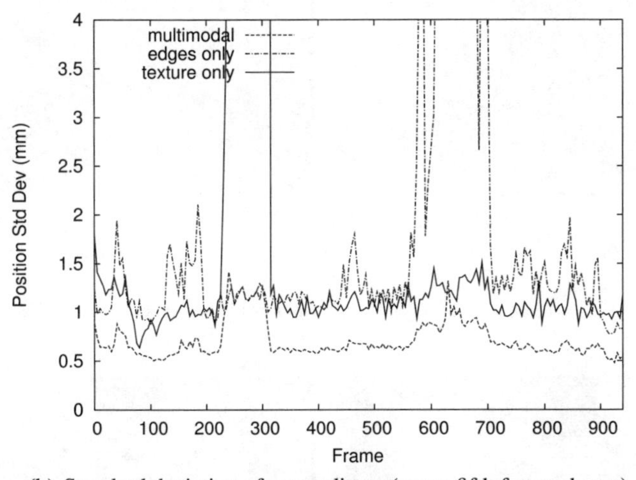

(b) Standard deviation of x-coordinate (every fifth frame shown)

Fig. 5.11. Performance of multi-cue tracker in occluded box tracking sequence.

of the edges causes a steady decrease in the number of associated features between frames 400 and 700, while the texture cues also decrease with the introduction of the blue box between frames 600 and 700 (see Figure 5.10(e)). Despite these losses, the multi-cue tracker robustly maintains the pose of the box through the entire sequence.

Figure 5.11(b) shows the standard deviation of the x-coordinate estimated using edge only, texture only and multi-cue trackers. As before, the multi-cue tracker achieves a smaller variance than the single-cue trackers, and the occlusion of edges and textures introduce large excursions in the error for the single-cue trackers. However, unlike the previous sequence, the single-cue trackers actually recover the pose of the object when the disturbances are removed. Since the target remains stationary during occlusion, the texture only tracker maintains a predicted pose that drifts only slowly from the actual pose, as shown in Figure 5.12(a). The accuracy of the predicted pose allows texture cues to be recovered once the object is removed (Figure 5.12(b)). When edges are later occluded, Figure 5.13(a) shows that the edge tracker develops association errors: the edges due to the occluding cups are mistaken for edges of the box, resulting in a pose bias. Figure 5.13(b) shows a similar error, in this case caused by the occluding hand. In both cases, the correct associations are eventually recovered after the occlusions are removed. Importantly, these association errors are not present in the output of the multi-cue tracker.

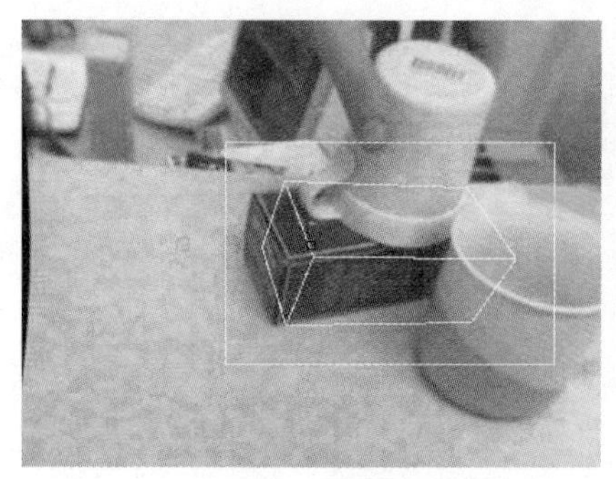

(a) Frame 300: drift in estimated pose due to occluded texture

(b) Frame 315: recovery of pose after reappearance of texture

Fig. 5.12. Performance of texture only tracker in the presence of occlusions.

(a) Frame 670: pose biased by distracting edges in occluding objects

(b) Frame 870: pose biased by distracting edges in occluding hand

Fig. 5.13. Performance of edge only tracker in the presence of occlusions.

5.5.3 Sequence 3: Rotation About Axis of Symmetry

This tracking sequence was designed to highlight a particular failure of edge-based tracking. In this case, a conical coffee mug is rotated anti-clockwise and then clockwise about its axis of symmetry (frames 0 to 250), and finally flipped upside-down (frames 250 to 500). As before, the sequence was processed using edge only, texture only and multi-cue trackers. Selected frames from the multi-cue tracking result are shown in Figure 5.14 (see *Multimedia Extensions* for the complete sequence). For each tracker, the angle between the initial and current pose of the mug is calculated from the angle/axis representation of the difference in orientation. Figure 5.15 shows the magnitude of the estimated change in angle for each tracker.

In this experiment, edge tracking fails to estimate the orientation of the mug due to insufficient constraints on the pose about the axis of symmetry. This is clearly demonstrated in the first 250 frames of Figure 5.15. While the other trackers closely follow the rotation of the mug during this period, the edge tracker simply obeys constant velocity dynamics and maintains the initial orientation. The rotation in the latter half of the sequence is successfully tracked by all cues since it does not involve an under-constrained degree of freedom. The issue of symmetry is not peculiar to edge-based tracking; an equivalent ambiguity arises for textures such as pin-stripes or concentric circles that also exhibit an unconstrained translational or rotational degree of freedom. Incidentally, texture only tracking fails in the final 100 frames of the sequence as the rotation of the mug causes a sufficiently large change in the appearance of textures. As in the previous results, only multi-cue tracking successfully estimates the pose of the object over the entire sequence.

5.5.4 Sequence 4: Ego-Motion Compensation

The final tracking sequence involves a stationary mug and a simple camera motion to demonstrate the pose bias introduced by ego-motion compensation. Figure 5.16(a) shows the scene when the mug was first scanned and modelled. The initial pose of the mug is overlaid as a black wireframe and the tracked pose (using multi-cue tracking) is shown in white. Obviously, the initial pose closely matches the target while both mug and cameras remain stationary. In Figures 5.16(b) and 5.16(c), the cameras are panned up and across the scene while the mug remains stationary. As described earlier, calibration errors in the neck joints will bias the estimated pose when transforming the object between the world frame and camera frame. This bias is clearly evident as a displacement between the initial pose (stored in the world frame) and actual target in both Figures 5.16(b) and 5.16(c). The bias observed during camera motion (Figures 5.16(b)) is considerably larger than the final bias due to the additional effect of latency in the joint angle readings. The close match between the tracked pose (white wireframe) and actual mug throughout the sequence demonstrates how multi-cue tracking overcomes the systematic error. This result highlights the necessity of object tracking even in static scenes for precision tasks such as grasping.

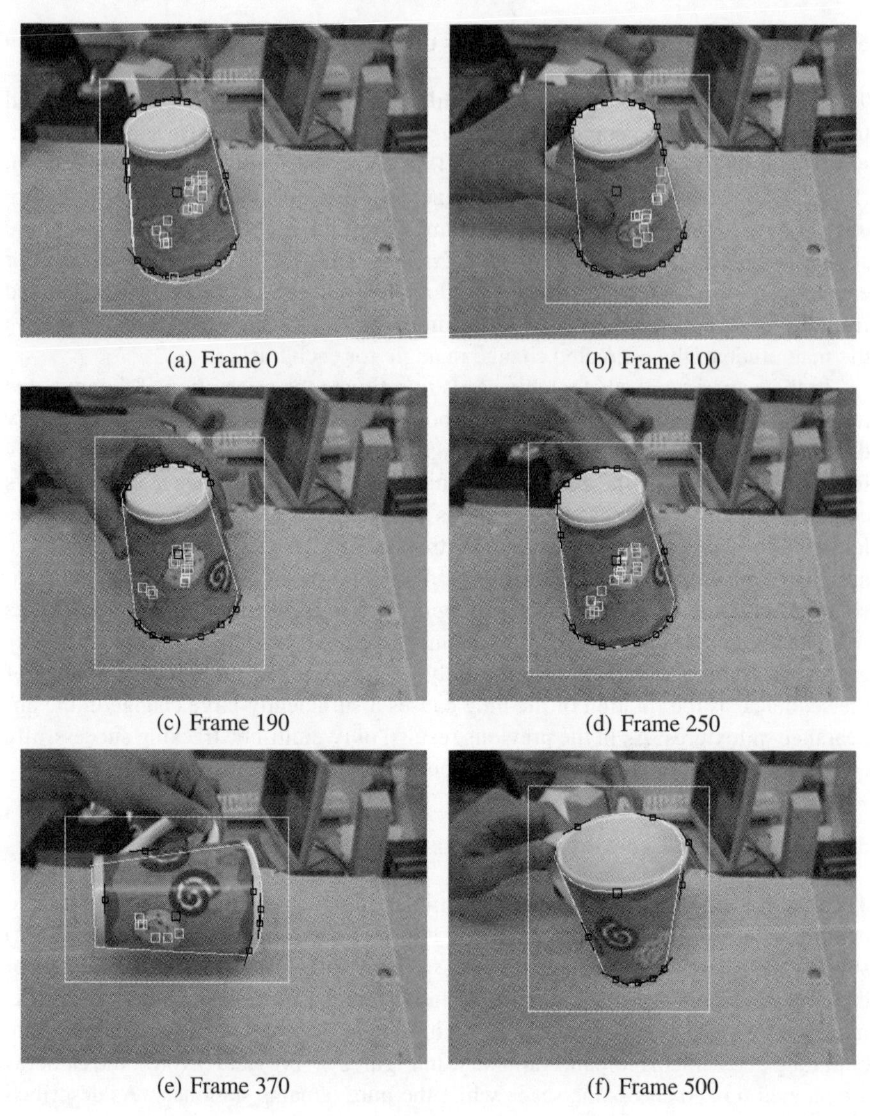

(a) Frame 0 (b) Frame 100

(c) Frame 190 (d) Frame 250

(e) Frame 370 (f) Frame 500

Fig. 5.14. Selected frames from coffee mug tracking sequence. Features used by multi-cue tracker are overlaid along with the estimated pose. Frame numbers are only approximate (see *Multimedia Extensions* for the full sequence).

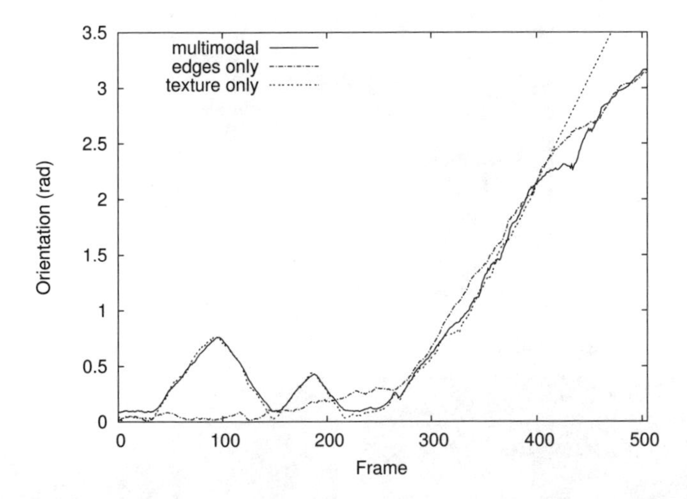

Fig. 5.15. Observed orientation of mug for single and multi-cue trackers.

5.6 Discussion and Conclusions

This chapter presented a multi-cue 3D model-based tracking algorithm that fuses colour centroid, intensity edge and texture template cues from stereo cameras in a Kalman filter framework. Cues are fused through the measurement model in the IEKF, which relate the observed features to the pose of the object. This approach allows stereo measurements to be used without explicit stereo correspondences, since all measurements interact through the filter state. Furthermore, the framework is completely extensible; fusion of additional cameras and cues simply requires a suitable measurement model.

Experimental results demonstrate that multi-cue fusion enables robust tracking in visual conditions that otherwise cause conventional single-cue tracking algorithms to fail, including low contrasting backgrounds, motion blur, lighting variations and occlusions. The robustness of multi-cue tracking relies on the constituent modalities having independent and complementary failure modes, so that a given visual distraction does not suppress all cues. Furthermore, the additional constraints provided by multi-cue tracking eliminate the problem of singularities in the estimated pose that arise for certain symmetries. The obvious direction for future work is to include additional modalities that have not yet been exploited. These might include new cues such as motion and depth, or functions of existing cues including curved intensity edges and higher order moments of colour blobs.

Calculation of the filter weight for optimal fusion requires accurate estimation of the measurement errors. The current implementation employs empirical values for the measurement errors, which are likely to produce a sub-optimal state estimate. This situation could be improved by estimating the observation error as part of the measurement process. In addition to providing a more accurate state covariance,

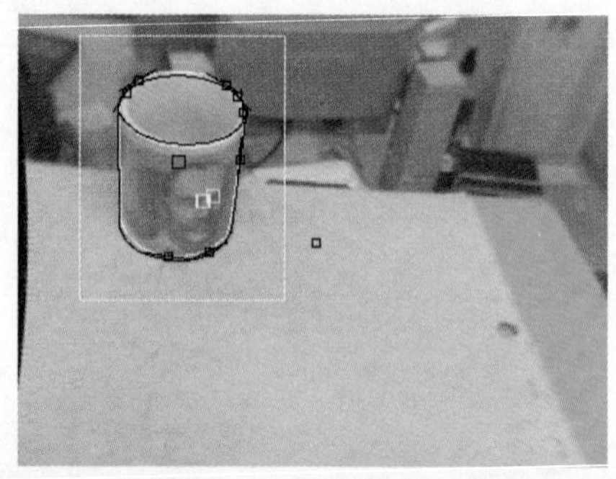

(a) Initial and tracked pose of the object closely match

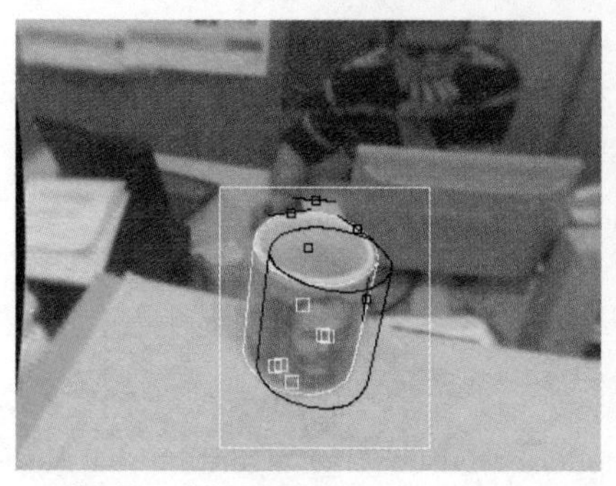

(b) Pose bias during camera motion

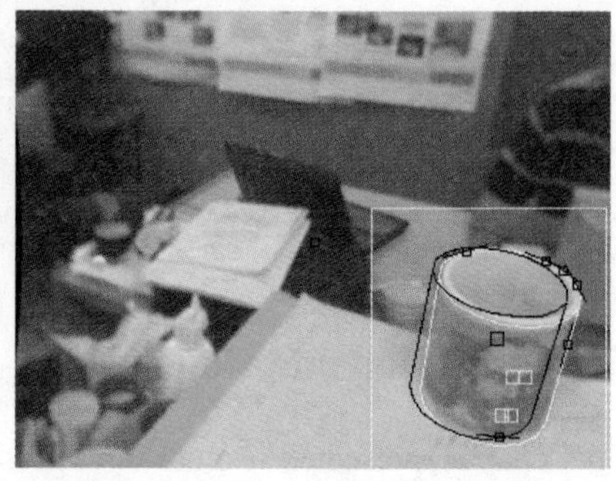

(c) Pose bias at final camera position

Fig. 5.16. Selected frames from ego-motion tracking sequence. Tracked pose is shown by the white wireframe model, and initial pose (compensated for ego-motion) is shown in black.

robustness would be improved by weighting the estimated pose towards the most reliable features under current conditions.

The measurement process could be further improved by using the estimated state covariance to guide the search for features. In the current implementation, measurements are restricted to a fixed ROI to reduce computational expense and eliminate background clutter. Unfortunately, a small fixed ROI also hinders recovery in the case of a tracking loss. By modulating the size of the ROI by the state covariance, efficient tracking could be maintained in favourable conditions while automatically increasing the search space when conditions degrade. The same applies to other thresholds used in edge and texture matching.

The current IEKF implementation assumes smooth motion through a constant velocity state transition model. This constraint is likely to be violated in an unstructured domestic environment where the dynamics are inherently unknown. Adaptive Kalman filtering (for example [164]) attempts to address this issue by estimating the state transition matrix, and may be a better suited to service applications. However, once an object is grasped, which is the ultimate aim in this work, the motion of the target becomes completely constrained by the robot. In this case, the constant velocity model should be replaced with a model describing the imposed motion.

Ultimately, Kalman filtering is only one of many tracking frameworks, both parametric and non-parametric. Particle filtering is another popular tracking framework in robotic applications, in which the Gaussian noise assumption is replaced by a sampled state distribution. Particle filters are generally more computationally expensive, but can also provide more robust recovery from tracking errors. The multi-cue approach developed in this chapter could be implemented with particle filtering among many other tracking frameworks.

6

Hybrid Position-Based Visual Servoing

The preceding chapters developed a framework for perception based on automatic extraction of 3D models from range data for object classification and tracking. In robotics, however, perception is only ever half the story! This chapter addresses the complementary problem of controlling a robotic manipulator to interact with the perceived world. Specifically, the controller must be able to drive an end-effector to some desired pose relative to a detected object. The traditional solution is kinematic control, in which joint angles form the control error and the pose of the end-effector pose is reconstructed through inverse kinematics. This approach can be effective for service robots when the camera parameters and kinematic model are well calibrated, as demonstrated in [10]. However, it is generally accepted that kinematic control deteriorates with increasing mechanical complexity [45]. Economic constraints impose additional limitations on the accuracy of calibration, including low manufacturing tolerances, cheap sensors and lightweight, compliant limbs for efficiency and safety. Achieving reliable, long term operation in an unpredictable environment reinforces the need to tolerate the effects of wear on sensors and mechanical components. Clearly, a more robust control solution is required.

Visual servoing is a robot control technique that minimizes the effect of calibration errors by using a camera to directly observe the relative pose between the object and end-effector. Section 2.4.2 reviewed contemporary visual servoing schemes, and this chapter shows how *position-based visual servoing* is a suitable control framework in the service robot domain. An extended approach, called *hybrid position-based visual servoing*, is proposed to overcome the limitations of traditional position-based visual servoing, namely a reliance on accurate camera calibration and loss of control when the end-effector is obscured. The hybrid controller dynamically estimates calibration-related parameters, and fuses visual tracking with kinematic measurements of the end-effector. This chapter also introduces the use of active visual cues to improve the robustness of end-effector tracking.

The following section motivates the proposed scheme by elaborating the challenges of visual servoing for service robots. Section 6.2 formalizes the servoing problem and describes the control law for position-based control. Sections 6.3 and 6.4 describe how the pose of the end-effector can be estimated from visual and kinematic measurements, including an analysis of calibration errors. Fusion of visual and kinematic measurements to estimate both pose and calibration parameters is discussed in Section 6.5. Section 6.6 describes image processing, including the use of active cues, and other implementation details. Finally, experimental results presented in Section 6.7 demonstrate the increased accuracy and robustness of hybrid position-based visual servoing compared to similar control techniques.

6.1 Introduction

To summarize the discussion in Section 2.4.2, visual servoing is an approach to robot control that uses visual measurements in the feedback loop. Visual servoing schemes are broadly classified as *image-based*, in which the control error is expressed in terms of features on the image plane, or *position-based*, in which the full pose of the robot and target are reconstructed before applying Cartesian control (see Figure 2.6). Various other approaches, including 2-1/2-D visual servoing [97] and affine approximations [24], have also been proposed to overcome the drawbacks of classical schemes. Visual servoing has been applied to a variety of domains, including mobile robotics [158], industrial manipulators [75] and unmanned aircraft [159]. Each application presents distinct challenges that motivate a unique formulation of the controller. It is therefore prudent to begin this chapter by considering the characteristics of service robots, the challenges of performing manipulation tasks in a domestic environment, and the influence of these factors on controller design.

Our first consideration is the configuration of the hand-eye system. As discussed in Section 2.4.2, visually servoed manipulators are usually configured as *eye-in-hand*, with the camera rigidly attached to the end-effector, or *fixed-camera*, with the camera and manipulator at opposite ends of a kinematic chain. The first configuration eliminates the problem of tracking the end-effector, if the hand-eye transformation is well calibrated. However, a fixed-camera offers greater flexibility to control the viewpoint and manipulator independently. This can be advantageous for service robots since the camera will likely serve many functions other than grasp control. For aesthetic reasons, fixed-camera is also the obvious configuration for humanoid service robots. The controller developed in this chapter is thus formulated for a fixed-camera hand-eye system. The application to an eye-in-hand implementation is straightforward and left as an exercise for the reader.

One of the attractions of image-based visual servoing is that the effect of camera calibration errors can be minimized by capturing and using an image of the robot in the desired pose as the controller reference [63, 65, 97]. Unfortunately, the singular, *ad hoc* nature of service tasks precludes any opportunity to directly observe the desired pose in advance. Instead, the controller reference must be predicted from an internal world model, which is constructed from the real world through an uncertain camera model. Uncertainties thus affect the positioning accuracy of the controller regardless of the selected visual servoing scheme, whether position-base or image-based. As will be shown later in the chapter, accurate position-based control can still be achieved by explicitly compensating camera calibration errors in the control loop.

Another perceived advantage of classical image-based control is that tracked features are easily constrained to remain within the field of view. Position-based schemes have also been developed to generate trajectories that maximize the availability of visual feedback [18]. However, the combination of *ad hoc* tasks and cluttered environments in service applications will likely introduce unavoidable occlusions that invalidate any advantage of these approaches. For example, a large initial pose error (when the end-effector is far from the target) may prevent the end-effector and target from being viewed simultaneously, and obstacles near the target may introduce occlusions when the end-effector should otherwise be visible. A control scheme for service robots must therefore emphasize robustness to occlusions rather than the impractical goal of maintaining continuous visual feedback. It will be shown that this is achieved by fusing visual and kinematic measurements in the feedback loop.

Collision avoidance is important for reliable operation in a cluttered domestic environment, and relies on the controller to generate predictable trajectories around obstacles in Cartesian space. Purely image-based control generates smooth image-plane trajectories but leads to unpredictable Cartesian motion and a higher likelihood of collisions. Conversely, position-based control, image-based control with decoupled orientation and translation [83, 129] and 2-1/2-D servoing [97] generate predictable Cartesian trajectories, and all exhibit similar stability and robustness [31]. Recent studies in applying biological reach-to-grasp strategies to robotic systems, which suggest that humans use 3D structural cues rather than image-based features [67], also hint at the importance of Cartesian control in human-like tasks. In fact, other results suggest that human motions are planned in Cartesian space rather than joint space [56], and that humans apply 3D interpretations to almost all visual information, including monocular images [60]. These observations are reflected in the emphasis on position-based visual servoing and other 3D techniques in the perception and control framework presented in this book.

In most visual servoing schemes, a kinematic model of the manipulator is ultimately required to transform the control error into joint velocities. The impact of kinematic calibration errors depends on the whether visual servoing is implemented as end open-loop (EOL) or end closed-loop (ECL) [68]. For a fixed camera configuration, EOL control is similar to the classical look-then-grasp approach in which a target object is tracked and the desired pose of the gripper is calculated through inverse kinematics. While EOL control simplifies the visual tracking problem, kinematic errors have a significant impact on positioning accuracy. Conversely, ECL control tracks both the end-effector and target, which significantly reduces the effect of kinematic uncertainties since the control error reduces to zero only when the end-effector is observed to reach the target. To further reduce the reliance on a kinematic model, some visual servoing schemes attempt to estimate the control error to joint velocity transformation directly [65, 110].

While robustness to kinematic calibration errors appears to make ECL control more suitable for service robots, this advantage is counterbalanced by a sensitivity to occlusions. That is, both the end-effector and target must be continuously visible to calculate the control error. Conversely, occlusions of the end-effector have no effect on the convergence of EOL control. To simultaneously provide robustness to calibration errors and occlusions, service robots would benefit from a *hybrid* approach borrowing aspects of both ECL and EOL visual servoing. This can be achieved by fusing kinematic and visual measurements to track the end-effector. The controller smoothly transits between visual and kinematic control as the observability of the end-effector varies. Fusion also improves the robustness of the tracker to visual distractions since kinematic measurements add further constraints on the pose of the end-effector. However, just as position-based ECL control is biased by camera calibration errors and EOL control is biased by kinematic errors, the hybrid approach must address both issues. As will be shown below, calibration errors can be compensated through on-line estimation of system parameters.

Yokokohji *et al.* [168] previously demonstrated fusion of kinematic and visual measurements for dextrous manipulation of an object using a three fingered hand. In that case, fusion was proposed to overcome the low resolution and sampling rate of visual control, while compensating for errors associated with a kinematic model describing the contact between the object and fingertips. Fusion was achieved through a least squares fit of the system state to the measurements, weighted by their respective errors. This approach is similar to the Kalman filter framework described below, but without fusing previous states or estimating the state covariance. Furthermore, the effects of occlusions and camera calibration errors were not considered. These deficiencies are addressed in the hybrid position-based visual servoing scheme described in the following sections.

6.2 Visual Servo Controller Overview

Figure 6.1 illustrates the relevant coordinate frames and transformations for a simple grasping task. In addition to the camera frame C and world frame W (introduced in Section 2.2.3), B is the base frame of the manipulator, E is the end-effector frame, object frame O describes the pose of the target, and the grasp frame G (rigidly attached to E) describes the desired position of the target relative to the gripper. This final frame is introduced to facilitate planning; various grasps are readily generated by setting the relative pose of G and E. For example, placing G between the fingertips of a hand-like gripper produces a *precision grasp*, while placing G near the base of the fingers results in a *power grasp*.

The basic task of visual servoing is to regulate the pose of the end-effector to align the grasp frame G with the object frame O. The ECL position-based controller described here is similar to the formulation in [68], with the addition of the grasp frame. The pose of the object and end-effector in the world frame (denoted by the transformations ${}^{W}\mathrm{H}_{O}$ and ${}^{W}\mathrm{H}_{E}$ in Figure 6.1) are estimated from visual measurements. The control error is then the pose error between the grasp and object frames (denoted as ${}^{G}\mathrm{H}_{O}$), which is identity when the frames are aligned at the completion of the control task. This control error is calculated from the measurements as

$$ {}^{G}\mathrm{H}_{O} = ({}^{W}\mathrm{H}_{E}\,{}^{E}\mathrm{H}_{G})^{-1} \cdot {}^{W}\mathrm{H}_{O} \tag{6.1} $$

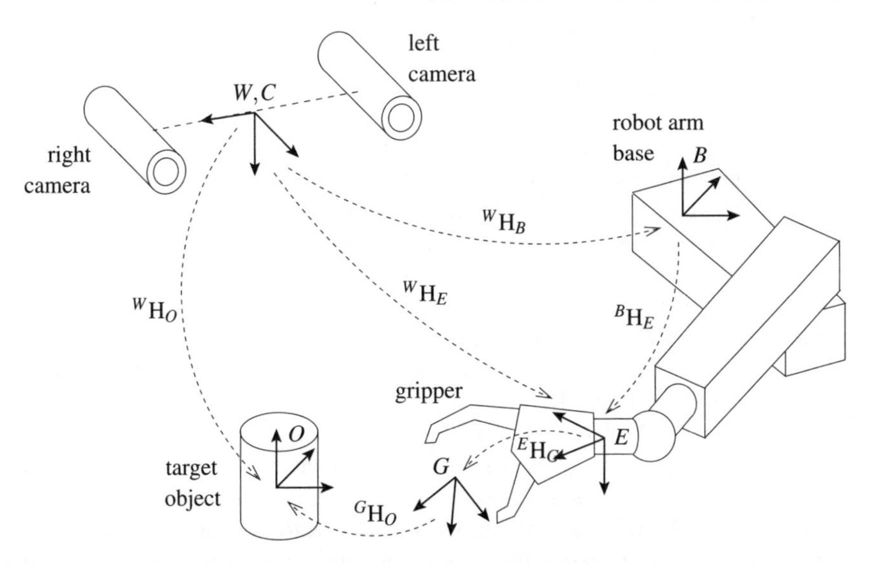

Fig. 6.1. Definition of visual servoing task with relevant coordinate frames and transformations. *Reprinted from [149]. ©2004, IEEE. Used with permission.*

where we recall that ${}^E\mathrm{H}_G$ is fixed in a given task. Expanding the rotational and translational components of ${}^G\mathrm{H}_O$ via equation (2.3) yields:

$$ {}^G\mathrm{R}_O = ({}^E\mathrm{R}_G {}^W\mathrm{R}_E)^{-1} {}^W\mathrm{R}_O \tag{6.2} $$

$$ {}^G\mathbf{T}_O = ({}^W\mathrm{R}_E {}^E\mathrm{R}_G)^{-1} [{}^W\mathbf{T}_O - {}^W\mathrm{R}_E {}^E\mathbf{T}_G - {}^W\mathbf{T}_E] \tag{6.3} $$

For each new observation of the object and end-effector, a proportional control law generates the required velocity screw of the end-effector to drive the pose error to identity. Denoting the rotational gain as k_1 and representing the orientation error ${}^G\mathrm{R}_O$ (equation (6.2)) as an angle of rotation ${}^G\theta_O$ about axis ${}^G\mathbf{A}_O$, the angular velocity component of the required velocity screw (expressed in the grasp frame) is

$$ {}^G\mathbf{\Omega} = k_1 {}^G\theta_O {}^G\mathbf{A}_O \tag{6.4} $$

The translational component of the velocity screw, using an independent gain of k_2, is

$$ {}^G\mathbf{V} = k_2 {}^G\mathbf{T}_O - {}^G\mathbf{\Omega} \times {}^G\mathbf{T}_O \tag{6.5} $$

where the first term drives the grasp frame towards the object and the second term compensates for the relative translational motion of the object induced by the angular velocity in equation (6.4). Transforming the above velocity screw to the robot base frame via equations (2.19)-(2.20) yields the velocity command sent to the robot:[1]

[1]The experimental implementation employs a Puma robot controlled in "tool-tip" mode; the desired velocity screw is actually specified in the end-effector frame, and the transformation to the robot frame in equations (6.6)-(6.7) is applied internally by the Puma controller.

$$^{B}\boldsymbol{\Omega} = {}^{B}\mathrm{R}_{E}{}^{E}\mathrm{R}_{G}{}^{G}\boldsymbol{\Omega} \tag{6.6}$$

$$^{B}\mathbf{V} = {}^{B}\mathrm{R}_{E}{}^{E}\mathrm{R}_{G}[{}^{G}\mathbf{V} - {}^{G}\boldsymbol{\Omega} \times ({}^{B}\mathrm{R}_{E}{}^{E}\mathbf{T}_{G} + {}^{B}\mathbf{T}_{E})] \tag{6.7}$$

To calculate the above command, the transformation ${}^{B}\mathrm{H}_{E}$ is reconstructed from measured joint angles and known kinematics. However, it is important to note that the velocity screw is zero when the pose error (equations (6.2)-(6.3)) is identity independently of ${}^{B}\mathrm{H}_{E}$. Thus, kinematic calibration errors in the calculation of ${}^{B}\mathrm{H}_{E}$ have no effect on the positioning accuracy of the controller (although dynamic performance may degrade). More importantly, the above formulation does not involve the transformation between the world frame and robot base frame (${}^{W}\mathrm{H}_{B}$ in Figure 6.1), meaning that the location of the robot need not be known if visual measurements are available (unlike EOL control).

The accuracy and robustness with which ${}^{W}\mathrm{H}_{E}$ and ${}^{W}\mathrm{H}_{O}$ (the pose of the end-effector and object) can be tracked imposes a fundamental limitation on the above controller. A framework for robustly tracking the object using texture, edge and colour cues was discussed in Chapter 5. For end-effector tracking, even greater accuracy and robustness to visual distractions and occlusions can be achieved by exploiting the fact that the gripper is part of the robot. This has two important consequences: firstly, we are free to design active, artificial visual cues that can be robustly detected, and secondly, kinematic measurements are also available to support visual tracking and enable calibration errors to be compensated. The following sections present a framework for fusing visual and kinematic measurements to robustly track the gripper. Section 6.3 presents measurement models and an error analysis for visual pose estimation, and Section 6.4 presents a similar discussion for kinematic measurements. Finally, a Kalman filter tracking framework to robustly estimate ${}^{W}\mathrm{H}_{E}$ by fusing visual and kinematic measurements is presented in Section 6.5.

6.3 Visual Feedback

6.3.1 Gripper Model with Active Visual Cues

This section describes a framework for model-based visual tracking using point cues at known locations on the end-effector. Artificial cues are commonly used in visual servoing, typically implemented as coloured dots or similar passive markers [68]. This approach offers simplicity and robustness in a controlled environment, but can be defeated in general applications where varying lighting conditions and background clutter distract passive feature detection. Avoiding these problems is a simple matter of replacing the passive cues with *actively lit* markers that maintain a constant appearance in the presence of lighting variations. Light emitting diodes (LEDs) provide a simple implementation of active cues, as shown in Figure 6.2(a). Another advantage of artificial cues is the ability to independently activate individual markers, which is exploited in Section 6.6) to solve association ambiguities.

Let the point set $\mathbf{G}_i$, $i = 1 \ldots n$, represent the locations of the n cues in the end-effector frame, which will be referred to as the gripper model. The gripper model for

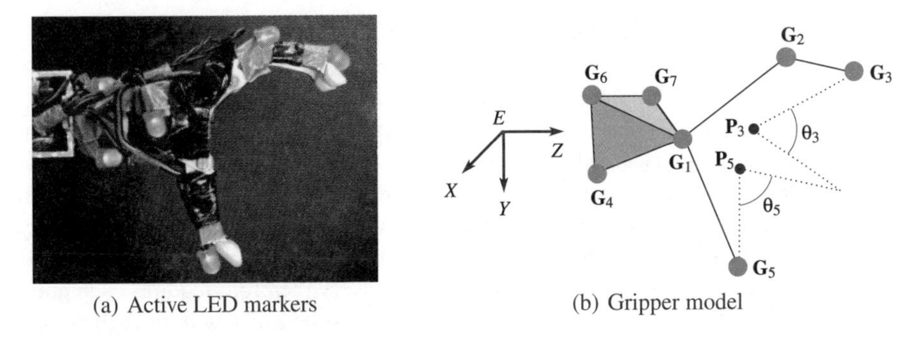

(a) Active LED markers (b) Gripper model

Fig. 6.2. Active LED features and articulated model for visual gripper tracking.

our experimental robot, with $n = 7$, is shown in Figure 6.2(b). The hand-like experimental gripper has a single DOF that allows the thumb and forefinger to open and close. The gripper model is similarly articulated so that $\mathbf{G}_5$, $\mathbf{G}_2$ and $\mathbf{G}_3$ reflect the current location of the finger cues. Let $\mathbf{P}_5$ and $\mathbf{P}_3$ represent the pivot points about which the thumb and forefinger rotate, and let θ_5 and θ_3 represent the maximum angles of rotation as shown in Figure 6.2(b). Furthermore, let $\widetilde{\mathbf{G}}_5$, $\widetilde{\mathbf{G}}_2$ and $\widetilde{\mathbf{G}}_3$ represent the nominal position of the finger cues for a fully open grasp, and let δ, where $0 \leq \delta \leq 1$, represent the fraction by which hand is closed, as measured by a joint encoder attached to the thumb. Then, the position of the finger cues are calculated as

$$\mathbf{G}_2 = \mathbf{P}_3 + \mathrm{R}_x(-\delta\theta_3)[\widetilde{\mathbf{G}}_2 - \mathbf{P}_3] \tag{6.8}$$

$$\mathbf{G}_3 = \mathbf{P}_3 + \mathrm{R}_x(-\delta\theta_3)[\widetilde{\mathbf{G}}_3 - \mathbf{P}_3] \tag{6.9}$$

$$\mathbf{G}_5 = \mathbf{P}_5 + \mathrm{R}_x(\delta\theta_5)[\widetilde{\mathbf{G}}_5 - \mathbf{P}_5] \tag{6.10}$$

where $\mathrm{R}_x(\theta)$ is the rotation matrix for a canonical rotation about the x-axis by angle θ. The gripper model, including opening angles and locations of LEDs and pivot points are manually calibrated for the experiments in Section 6.7.

6.3.2 Pose Estimation

In a 3D model-based tracking framework, the pose of the gripper is estimated by matching the gripper model to stereo image plane measurements of the artificial cues. In the following discussion, let the measurements of the n markers on the left and right image plane be represented by ${}^L\mathbf{g}_i$ and ${}^R\mathbf{g}_i$, $i = 1 \dots n$, respectively.

The matching and estimation problem can be solved using two distinct approaches. The first approach is to reconstruct corresponding stereo measurements of individual markers before fitting the gripper model to the resulting set of 3D points. The advantage of this approach is that stereo triangulation and model-based pose recovery from point features are well studied problems with closed form solutions [48, 54]. Furthermore, this approach leads to a linear measurement function in the tracking filter since the pose forms both the measurement and state. However, this method also has several important drawbacks. Firstly, corresponding stereo measurements of each point may not always be available due to occlusions (including self-occlusions). When a marker is visible in only one camera, potentially useful information simply discarded. Furthermore, the recovered pose will not be optimal with respect to the measurement error if the points are uniformly weighted for model fitting. When 3D points are reconstructed from stereo measurements, the uncertainty is greatest in the Z-direction and increases for distant points as the reconstruction becomes less well-conditioned. Optimal pose recovery therefore requires each point to be weighted by a non-uniform covariance that must be calculated for each point.

The second approach is to minimize the *reprojection error* on the image plane between the measured markers and the projection of the gripper model, such that the projected model point appear closest to the measurements at the optimal estimated pose. This method leads to a more complex tracking filter since the measurement equations involved a non-linear camera projection, and minimization of the image plane error requires a numerical solution. However, this approach has several distinct advantages over the method described above. Firstly, a uniform error variance can be assumed over all measurements, since the error function is formed on the image plane. More importantly, this approach does not require explicit stereo correspondences; rather, each measurement is only matched to the projection of a point in the gripper model. Thus, all available measurements contribute to the estimated pose without discarding markers visible in only one camera. In fact, this method extends to a general n-camera configuration by including an appropriate projection for each camera. A monocular version is used for position-based visual servoing in [163], and the following sections describe a stereo implementation.

Let $\mathbf{p}_E$ represent the gripper pose in the world frame and $\mathrm{H}(\mathbf{p}_E) = {}^W\mathrm{H}_E$ represent the equivalent homogeneous transformation matrix. Furthermore, let ${}^C\mathrm{H}_W$ represent the transformation from the world frame to the camera frame (due to the active pan/tilt head), which is the inverse of the transformation in equation (2.30), and let ${}^{L,R}\mathrm{P}$ represent the left and right camera projection matrices given by equation (2.29). Now, transforming the 3D model points $\mathbf{G}_i$ from the end-effector frame to the camera frame and projecting onto the stereo image planes gives the predicted measurements ${}^{L,R}\hat{\mathbf{g}}_i(\mathbf{p}_E)$ for the given pose $\mathbf{p}_E$:

$${}^{L,R}\hat{\mathbf{g}}_i(\mathbf{p}_E) = {}^{L,R}\mathrm{P}\,{}^C\mathrm{H}_W\mathrm{H}(\mathbf{p}_E)\mathbf{G}_i \tag{6.11}$$

The optimal estimate of the pose, denoted by $\hat{\mathbf{p}}_E$, minimizes the reprojection error between the predicted and actual measurements. The reprojection error $D^2(\mathbf{p}_E)$ for a given pose $\mathbf{p}_E$ is the sum of the squared distances between matching points given by

$$D^2(\mathbf{p}_E) = \sum_i d^2({}^L\hat{\mathbf{g}}_i(\mathbf{p}_E), {}^L\mathbf{g}_i) + d^2({}^R\hat{\mathbf{g}}_i(\mathbf{p}_E), {}^R\mathbf{g}_i) \tag{6.12}$$

where ${}^{L,R}\mathbf{g}_i$ are the actual measurements, ${}^{L,R}\hat{\mathbf{g}}_i$ are given by equation (6.11), and $d(\mathbf{x}_1, \mathbf{x}_2)$ defines the Euclidean distance between $\mathbf{x}_1$ and $\mathbf{x}_2$. The estimated pose is thus recovered by the minimization

$$\hat{\mathbf{p}}_E = \text{argmin}[D^2(\mathbf{p}_E)] \tag{6.13}$$

The equal weighting of all measurements to form the reprojection error in equation (6.12) is consistent with the assumption of uniform error variance for all image plane measurements, and the estimated pose is optimal with respect to measurement error.

6.3.3 Compensation of Calibration Errors

Any real vision system will suffer some uncertainty in the estimated intrinsic and extrinsic camera parameters (in addition to the inherent approximations in the camera model). At best, the pose of the end-effector recovered from equations (6.12)-(6.13) is some unknown transformation of the real pose. The following discussion develops an error model to analyse the impact of uncertainties in camera calibration on the accuracy of the estimated pose. This analysis will also reveal strategies to compensate for calibration errors during visual servoing. A complete derivation of the error model is given in Appendix D.

We begin this discussion by considering the reconstruction of a single point $\mathbf{X}$ from corresponding stereo measurements ${}^L\mathbf{x} = ({}^Lx, {}^Ly)^\top$ and ${}^R\mathbf{x} = ({}^Rx, {}^Ry)^\top$. Let $\widehat{\mathbf{X}}$ represent the estimated 3D position recovered from minimization of the reprojection error (equations (6.12)-(6.13)) for a single point. It is easily shown (see Section D.1) that the optimal reconstruction is

$$\widehat{\mathbf{X}} = \frac{b}{{}^Lx - {}^Rx}\left({}^Lx + {}^Rx,\ {}^Ly + {}^Ry,\ 2f\right)^\top \tag{6.14}$$

Now, let b^*, f^* and ν^* represent the *actual* baseline, focal length and verge angle of the stereo rig, and let b, f and ν represent the *calibrated* parameters of the camera model. The task is to determine how the deviation between actual and calibrated parameters affects the reconstruction in equation (6.14). We first note that the verge angle is used to projectively rectify non-rectilinear image plane measurements as described in Section 2.2.4. When the estimated verge angle differs from the actual verge angle ($\nu - \nu^* \neq 0$), projective rectification over-compensates by $(\nu - \nu^*)$. In this case, the ideal rectilinear stereo pin-hole projection relating $\mathbf{X}$ and ${}^{L,R}\mathbf{x}$ given by

$${}^{L,R}\mathbf{x} = \frac{f^*}{Z}(X \pm 2b^*, Y)^\top \tag{6.15}$$

is replaced by the over-corrected model (see Section D.2)

$${}^{L,R}\mathbf{x} = \frac{f^*\left((X \pm 2b^*)\cos(\nu - \nu^*) \mp Z\sin(\nu - \nu^*),\ Y\right)^\top}{Z\cos(\nu - \nu^*) \pm (X \pm 2b^*)\sin(\nu - \nu^*)} \tag{6.16}$$

where the top sign is taken for L and the bottom for R. The analysis proceeds by applying a small angle approximation to equation (6.16) (assuming a mild verge error, $\nu - \nu^* \ll 1$) and substituting the result into the reconstruction equation (6.14). Taking a Taylor series expansion with respect to f, b and ν about the operating point $f = f^*$, $b = b^*$ and $\nu = \nu^*$, it then follows that the relationship between the actual point $\mathbf{X}$, and point reconstructed via an imperfect camera model $\widehat{\mathbf{X}}$ can be approximated by (see Section D.2):

$$\widehat{\mathbf{X}} \approx \left(1+\frac{b-b^*}{2b^*}+\frac{X^2+Z^2}{2Zb^*}(\nu-\nu^*)\right)\begin{pmatrix}X\\Y\\Z\end{pmatrix}+\frac{f-f^*}{f^*}\begin{pmatrix}0\\0\\Z\end{pmatrix}+(\nu-\nu^*)\begin{pmatrix}2Xb^*/Z\\0\\2b^*\end{pmatrix} \tag{6.17}$$

Finally, we note that the distances between gripper features are generally much smaller than the distance from the camera to the gripper. In this case, the following ratios in equation (6.17) can be approximated by the constants k_1 and k_2:

$$\frac{X^2+Z^2}{2b^*Z} \sim k_1, \quad \frac{2b^*X}{Z} \sim k_2 \tag{6.18}$$

The relationship between $\mathbf{X}$ and $\widehat{\mathbf{X}}$ then reduces to a linear function of b, f and ν:

$$\widehat{\mathbf{X}}(b,f,\nu) \approx K_1(b,\nu)\mathbf{X}+K_2(f)(0,\,0,\,Z_i)^\top+\mathbf{T}(\nu) \tag{6.19}$$

where

$$K_1(b,\nu) = 1+\frac{b-b^*}{2b^*}+k_1(\nu-\nu^*) \tag{6.20}$$

$$K_2(f) = \frac{f-f^*}{f^*} \tag{6.21}$$

$$\mathbf{T}(\nu) = (\nu-\nu^*)(k_2,\,0,\,2b^*)^\top \tag{6.22}$$

Equation (6.19) reveals that a verge angle error translates the reconstructed point by $\mathbf{T}(\nu)$ relative to $\mathbf{X}$. Importantly, this bias has negligible effect on the accuracy of ECL visual servoing. Since the object and gripper will suffer approximately equal bias when aligned, the *relative* pose error remains unbiased. The focal length error scales the reconstructed point in the Z direction by $K_2(f)$, which can be expected from equation (6.14) where f appears in only the Z term. While this scaling can affect model-based tracking, focal length is typically the most accurately known or readily calibrated camera parameter. Thus, this source of error will also be neglected in the remaining discussion. The final source of error is a scaling of the reconstruction by $K_1(b,\nu)$ due to baseline and verge angle error. This is to be expected for the baseline, since b is a factor of $\widehat{\mathbf{X}}$ in equation (6.14). Accurate calibration of the verge angle and baseline require the cameras to be precisely mounted on the active pan/tilt head, which is difficult in practice. Furthermore, the baseline may change dynamically if the camera centres are not aligned with the verge axes. Thus, we propose that the main contribution to error in the reconstruction is the scale term $K_1(b,\nu)$.

The above analysis involved only a single point $\mathbf{X}$, and we now consider the effect of the scale error $K_1(b,\nu)$ on the estimated pose of an object represented by multiple points. Intuition leads to the conclusion that scaling the reconstructed points by $K_1(b,\nu)$ accordingly scales the estimated pose. If this were true, visual servoing would still converge since the pose error would still be zero when the (scaled) gripper and object are aligned. However, it is shown below that $K_1(b,\nu)$ does *not* lead to a simple scale of the estimated pose.

The following analysis is simplified by considering only pure translation of the gripper, represented by $\mathbf{T}_E = (X_E, Y_E, Z_E)^\top$. As before, let K_1 represent the scale due to camera calibration errors and let $\mathbf{G}_i = (X_i, Y_i, Z_i)^\top$, $i = 1 \ldots n$, represent the actual locations of n markers in the gripper frame. Without loss of generality, the mean position of markers in the gripper frame is assumed to be zero ($\sum \mathbf{G}_i = \mathbf{0}$). Now, let $\mathbf{G}_i^*$ represent the (inaccurately) measured location of markers in the world frame, that is, after scaling by K_1 due to calibration errors. The locations of these features can be expressed as

$$\mathbf{G}_i^* = K_1(\mathbf{G}_i + \mathbf{T}_E) \tag{6.23}$$

The actual measurements ${}^{L,R}\mathbf{g}_i$ are the projection of these points onto the left and right image planes (taking the top sign for L and bottom for R), given by:

$${}^{L,R}\mathbf{g}_i = \frac{f}{K_1(Z_i + Z_E)}(K_1(X_i + X_E) \pm 2b,\ K_1(Y_i + Y_E))^\top \tag{6.24}$$

Now, let $\widehat{\mathbf{T}}_E = (\widehat{X}_E, \widehat{Y}_E, \widehat{Z}_E)^\top$ represent the estimated pose of the gripper by minimization of the reprojection error (equations (6.12)-(6.13)). In calculating the reprojection error, the *unscaled* model is translated by $\widehat{\mathbf{T}}_E$ and projected onto the image plane to give the predicted measurements:

$${}^{L,R}\hat{\mathbf{g}}_i = \frac{f}{Z_i + \widehat{Z}_E}(X_i + \widehat{X}_E \pm 2b,\ Y_i + \widehat{Y}_E)^\top \tag{6.25}$$

The reprojection error between the predicted and actual measurements is calculated by substituting equations (6.24) and (6.25) into equation (6.12). Solving equation (6.13) analytically, it can be shown (see Section D.3) that the optimal estimate of the gripper translation minimizing the reprojection error is given by:

$$\widehat{\mathbf{T}}_E = \frac{X_E \sum X_i Z_i + Y_E \sum Y_i Z_i - Z_E(\sum X_i^2 + \sum Y_i^2 + 4nb^2)}{K_1(X_E \sum X_i Z_i + Y_E \sum Y_i Z_i) - Z_E[K_1(\sum X_i^2 + \sum Y_i^2) + 4nb^2]} K_1 \mathbf{T}_E \tag{6.26}$$

The above equation shows that the intuitive result $\widehat{\mathbf{T}}_E = K_1 \mathbf{T}_E$ (ie. the estimated pose is simply scaled by K_1) only occurs when $n = 1$ (a single point) or $K_1 = 1$ (perfect calibration). Otherwise, the pose bias caused by camera calibration errors is a function of the arrangement of points in the model itself. In other words, the estimated pose for two different objects at the *same* position, in the presence of the *same* calibration errors, will be different! This result can be understood by considering the effect of unknown scale in monocular images; large objects appear closer and small objects further away, even though they may be equally distant from the camera[2]. This has a serious and detrimental effect on visual servoing, since the pose error may not converge to zero when the object and gripper frames are aligned due to a differing bias in the estimated pose.

[2] A well-used effect in TV and movie production known as "forced perspective".

The proposed solution to compensate for calibration errors is to estimate both the pose $\mathbf{p}_E$ and unknown scale K_1 of the gripper in minimizing the reprojection error. The predicted measurements in equation (6.11) are replaced with

$$^{L,R}\hat{\mathbf{g}}_i(K_1,\mathbf{p}_E) = {}^{L,R}\mathrm{P}^C\mathrm{H}_W\mathrm{H}(\mathbf{p}_E)\cdot(K_1{}^E\mathbf{G}_i) \tag{6.27}$$

The pose is estimated as before, by substituting the predicted and actual measurements into equation (6.12) and minimizing the observation error with respect to the estimated parameters. In this case, the minimization is:

$$(K_1,\hat{\mathbf{p}}_E) = \mathrm{argmin}[D^2(K_1,\mathbf{p}_E)] \tag{6.28}$$

To sufficiently constrain the scale, four or more measurements are required with at least one feature on each image plane. Monocular measurements do not sufficiently constrain the scale since the projective transformation in equation (2.23) is only defined up to an unknown scale (ie. central projection is a many-to-one mapping). Section 6.5 describes how unconstrained parameters are handled in the implementation of the tracking filter. Finally, any remaining bias due to unmodelled camera errors are assumed to be sufficiently small to be neglected, which is verified experimentally in Section 6.7.

6.4 Kinematic Feedback

At the most basic level, kinematic feedback originates from the angle encoders attached to the joints of the robot. As with the visual pose estimation, the kinematic pose of the gripper is estimated by minimizing the error between the predicted measurements (as a function of pose) and the actual measurements. Rather than formulating the error directly in terms of joint angles, the pose of the gripper is first recovered via a kinematic model that encodes the transformations between the links of the manipulator (see [102] for a discussion of robot kinematics), and the error is expressed in terms of pose vectors. Let $\boldsymbol{\theta}$ represent the joint angle vector, and let ${}^B\mathrm{H}_E(\boldsymbol{\theta})$ represent the measured gripper pose (with respect to the robot base) obtained from the kinematic model. Now, let ${}^B\widehat{\mathrm{H}}_E$ represent the predicted measurement, given by

$${}^B\widehat{\mathrm{H}}_E = {}^W\mathrm{H}_B^{-1}\cdot{}^W\widehat{\mathrm{H}}_E \tag{6.29}$$

Section 6.5 describes how the measurement error is calculated from the equivalent 6-dimensional pose vectors representing the transformations ${}^B\widehat{\mathrm{H}}_E$ and ${}^B\mathrm{H}_E(\boldsymbol{\theta})$.

The predicted measurement in equation (6.29) requires knowledge of ${}^W\mathrm{H}_B$, the location of the robot base in the world frame[3]. Many methods for calibrating this transformation are reported in the literature, including the use of special calibration targets [156] and observation of known robot motions [6, 61]. Typically, the robot position is calibrated once during initialization and thereafter assumed to remain

[3]This is equivalent to the so-called *hand-eye* transformation for eye-in-hand systems.

constant. However, this approach suffers from several difficulties in the presence of calibration errors. In particular, the visual measurements used for calibration may be biased by camera calibration errors, while compliant/flexible manipulators may produce dynamic and unpredictable deviations from the kinematic model. In the latter case, static calibration will result in biased kinematic measurements even if the robot position is well known.

A solution to this problem is to treat the robot position as a dynamic bias between the measured kinematic pose (in the robot base frame) and the pose estimated from visual measurements (in the world frame). This is achieved by adding the pose parameters for ${}^{B}\mathrm{H}_{W}$ to the state vector of the tracking filter. For each new set of kinematic and visual measurements, ${}^{B}\mathrm{H}_{W}$ is updated through the measurement function in equation (6.29), providing continuous, on-line calibration. Solving this unknown transformation requires the pose of the gripper to be sufficiently constrained by both kinematic and visual measurements; kinematic pose is always available, while visual pose recovery requires three or more image plane measurements as described above. When the solution is unconstrained, ${}^{B}\mathrm{H}_{W}$ does not updated in the tracking filter, as described in the next section.

6.5 Fusion of Visual and Kinematic Feedback with Self-Calibration

As described in Section 6.2, position-based visual servoing requires accurate and robust tracking of the end-effector. This is achieved through two mechanisms: fusion of the visual and kinematic pose estimators described in the previous sections, and on-line estimation of calibration parameters. Kinematic feedback allows servoing to continue when the gripper is occluded or outside the field of view, while visual feedback improves positioning accuracy by observing the gripper and target directly. The visual and kinematic estimators are both biased by calibration errors, which can be compensated by estimating a few simple calibration parameters.

The *Iterated Extended Kalman filter* (IEKF) was introduced in Chapter 5 for multi-cue model-based target tracking, and is also ideal for tracking the gripper from visual and kinematic measurements. The Kalman filter is a well-known algorithm for optimally estimating the state of a linear dynamic system from noisy measurements, and the IEKF extends the framework to non-linear systems by solving the filter equations numerically. This framework is used in many visual and robotic tracking applications, including visual servoing (for example [163]). Appendix C describes the basic equations of the IEKF, and a detailed treatment of Kalman filter theory can be found in [8, 74].

The Kalman filter is a statistically robust framework for sensor fusion, which is exploited here to fuse visual and kinematic measurements. Let $\mathbf{x}(k)$ represent the state (the parameters we wish to estimate) of the IEKF at time k. Following the discussion in the Sections 6.3 and 6.4, the state can be summarized as

$$\mathbf{x}(k) = (\mathbf{p}_E(k), \dot{\mathbf{r}}_E(k), {}^{B}\mathbf{p}_W(k), K_1(k))^\top \tag{6.30}$$

where $\mathbf{p}_E(k)$ is the pose of the end-effector and $\dot{\mathbf{r}}_E(k)$ is the velocity screw of the end-effector (both in the world frame), ${}^B\mathbf{p}_W(k)$ is the pose vector representing the inverse of ${}^W\mathrm{H}_B$ (the location of the robot in the world frame, used in equation (6.29)), and $K_1(k)$ is the scale parameter for camera calibration (used in equation (6.27)).

Now, let $\hat{\mathbf{x}}(k)$ represent the state estimated by the IEKF at time k. At the next time step, a new state $\hat{\mathbf{x}}(k+1)$ is estimated using a two stage process: *state prediction* from a known dynamic model, followed by *state update* based on the new measurements. For the first step, the gripper pose is modelled by the familiar constant velocity dynamics (assuming a smooth trajectory), while the calibration parameters are modelled as static values. Thus, the predicted state variables at time $k+1$ based on the estimated state at time k and the dynamic models are

$$\mathbf{p}_E(k+1) = \mathbf{p}_E(k) + \Delta t \dot{\mathbf{r}}_E(k) \tag{6.31}$$

$$\dot{\mathbf{r}}_E(k+1) = \dot{\mathbf{r}}_E(k) \tag{6.32}$$

$${}^B\mathbf{p}_W(k+1) = {}^B\mathbf{p}_W(k) \tag{6.33}$$

$$K_1(k+1) = K_1(k) \tag{6.34}$$

where Δt is the sample time between state updates. The IEKF also requires an estimate of the state transition noise ($\mathbf{v}(k)$ in equation (C.1), also known as the process noise), to correctly weight predicted state against the measurements in the update step. For this purpose, it is sufficient to assume the state variables have independent noise with fixed variance.

The state update step then estimates the new state as a weighted average of the predicted state and the state estimated from the new measurements at time $k+1$. These are represented by the vector $\mathbf{y}(k+1)$ given by

$$\mathbf{y}(k+1) = ({}^L\mathbf{g}_1(k+1), {}^R\mathbf{g}_0(k=1), \dots, {}^R\mathbf{g}_n(k+1), {}^B\mathbf{p}_E(k+1))^\top \tag{6.35}$$

where ${}^{L,R}\mathbf{g}_i(k+1)$, $i = 1 \dots n$, are the measured locations of the n active markers on the stereo image planes, and ${}^B\mathbf{p}_E(k+1)$ is the gripper pose (in the robot base frame) reconstructed from joint angles and the known kinematic model. As described earlier, the state is estimated from the measurements by minimizing the observation error between the predicted and actual measurements. The prediction models are provided by equation (6.27) for visual measurements and (6.29) for kinematic measurements. These models are the mechanism by which the IEKF achieves multimodal fusion, by coupling the visual and kinematic measurements to a common underlying state.

To correctly weight the components of the observation error, an estimate of the measurement noise ($\mathbf{w}(k+1)$ in equation (C.3)) is required. As with the process noise, we can assume the measurements have independent error with fixed variance. However, the tracking filter must also cope with the possibility that some markers are occluded and produce no valid measurement. In Chapter 5, this problem was solved using a variable dimension measurement vector that represented only visible cues. An alternative approach used in this chapter is to maintain a fixed dimension measurement vector and set a very large error variance (low confidence) for occluded markers, which are thus associated with negligible weight in the state update.

As described in Chapter 5, care must be taken when dealing with angular state variables to avoid the problems of non-uniqueness and degeneracy in the state update (see also Section 2.1.3). Again, the approach of [15, 161] successfully avoids these problems. The orientation parameters of ${}^{W}\mathbf{p}_E$ and ${}^{B}\mathbf{p}_W$ are represented as differential Euler angles in the state vector, with the total orientations stored outside the state vector as quaternions. The transformations ${}^{W}\widehat{\mathrm{H}}_E$ and ${}^{B}\mathrm{H}_W$ in equations (6.27) and (6.29) are therefore calculated in the following form:

$$\mathrm{H} = \begin{pmatrix} \mathrm{R}(\mathbf{q})\mathrm{R}(\Delta\phi, \Delta\theta, \Delta\psi) & \mathbf{T} \\ \mathbf{0}_{1\times 3} & 1 \end{pmatrix} \tag{6.36}$$

where $\mathbf{T}$ is the translational component of the pose, $\mathbf{q}$ is the quaternion representing the total orientation, $\Delta\phi$, $\Delta\theta$ and $\Delta\psi$ are the differential Euler angles from the state vector and $\mathrm{R}(\cdot)$ is the rotation matrix representing the given orientation parameters. For each state update, the incremental Euler angles are integrated into the external quaternions using equations (2.7) and (2.12), and then set to zero in the state vector before the next update.

When dealing with angular measurements, care must also be taken to ensure the observation error is uniquely defined (for example, the error between 0 and 2π radians is $2\pi n$ for any integer n). Fortunately, representing the orientation of ${}^{B}\mathbf{p}_E$ by a quaternion provides the necessary uniqueness. However, the quaternion components cannot be assumed to have independent and fixed error variance due to the constraint imposed by normalization (see Section 2.1.3). This problem is resolved by assuming independent errors for the equivalent Euler angles. The Euler angle variances are then transformed to the quaternion covariance by linearizing the relationship between Euler angles and quaternions (equation (2.7)).

As mentioned in the previous sections, special care is taken to ensure the calibration parameters in the estimated state are sufficiently constrained. It is well known that the pose of a point-based model can be recovered from three monocular measurements, although there may be up to four possible solutions [49]. However, three points may be insufficient for stereo cameras if two measurements correspond to the same point on different image planes (providing only five constraints instead of six). Furthermore, K_1 is unconstrained for monocular visual measurements, while ${}^{B}\mathbf{p}_W(k)$ requires the gripper pose to be constrained by visual measurements. Finally, avoiding association errors in marker detection requires support across multiple consistent measurements (see Section 6.6.1). To simplify the problem of constraining the estimated state, the following hierarchy of estimators is adopted based on the number of measured features, n_L and n_R, on each image plane:

$n_L \geq 3$ and $n_R \geq 3$: Sufficient visual measurements are available to minimize the likelihood of association errors and constrain all parameters of the state vector.

$n_L \geq 3$ or $n_R \geq 3$ (but not both): With fewer than three measurements, all measurements on the given image plane are discarded due to possible association errors. The monocular measurements on the remaining image plane are sufficient to constrain the visual pose of the end-effector and robot location. However, the scale is unconstrained and thus excluded from the state update.

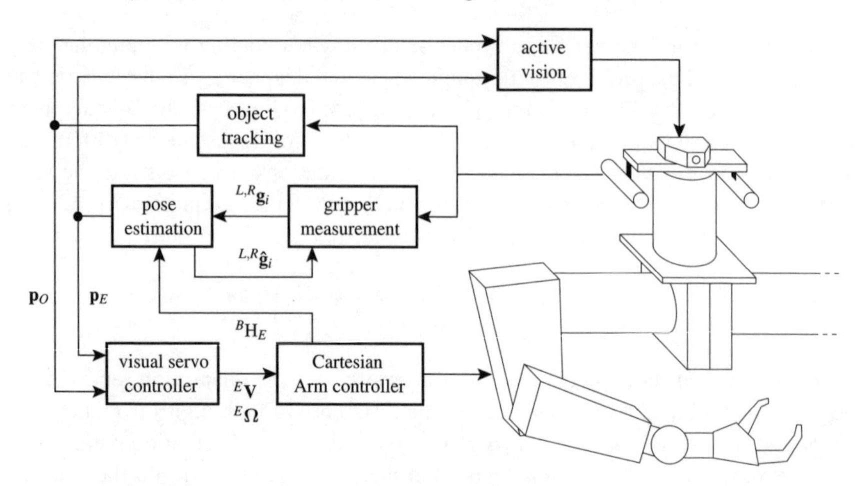

Fig. 6.3. Block diagram of hybrid position-based visual servoing control loop. *Reprinted from [149]. ©2004, IEEE. Used with permission.*

$n_L < 3$ and $n_R < 3$: All visual measurements are discarded due to possible association errors. The end-effector is estimated using only kinematic measurements, and all calibration parameters are excluded from the state update.

When necessary, K_1 or ${}^B\mathbf{p}_W(k)$ are excluded from the state update by setting the corresponding rows and columns of the measurement Jacobian (see equation (C.8)) to zero. The experimental results in Section 6.7.2 demonstrate the successful operation of the tracking filter under all of the above conditions.

The operation of the IEKF can now be summarized as follows. At each time step, the filter first predicts the current state using the dynamic model in equations (6.31)-(6.34). The predicted state, which includes both the pose of the gripper $\mathbf{p}_E$ and the calibration parameters ${}^B\mathbf{p}_W$ and K_1, is used to predict the visual and kinematic measurements via equations (6.27) and (6.29). The observation error is then formed by taking the difference between the predicted and actual measurements. The new state is estimated by averaging the predicted state and observation error weighted by the Kalman gain (calculated from the state and measurement covariances using the Kalman filter update equations). As described above, care is taken to exclude the calibration parameters from the state update if insufficient visual measurements are available. Finally, the differential orientation of $\mathbf{p}_E$ and ${}^B\mathbf{p}_W$ in the state vector are integrated into external quaternions representing total orientation, and reset to zero for the next time step. Appendix C details the IEKF equations in each of these steps.

6.6 Implementation

This section presents the implementation of hybrid position-based visual servoing on our experimental hand-eye platform (see Figure 7.1). A block diagram of the implementation is shown in Figure 6.3. The *active vision* block controls the camera view to maximize visual feedback using the proportional control laws in equations (2.31)-(2.33). During servoing, different active vision strategies are employed depending on the pose error. When the object and gripper cannot be observed simultaneously due to large separation, active vision tracks only the object while visual servoing relies on EOL control. Once the end-effector nears the object, active vision focuses on the mid-point between the object and gripper to ensure both remain observable.

The visual servoing control loop is divided into three sets of blocks: measurement, state estimation and actuation. The *gripper measurement* block detects the markers using the process described below in Section 6.6.1. The *pose estimation* block then estimates state of the IEKF presented in Section 6.5 and feeds the predicted gripper pose back to the measurement block to aid future detections. Similarly, the *object tracking* block estimates the pose of the object as described in Chapter 5. The *visual servo controller* then calculates the relative pose error and desired velocity screw (see Section 6.2), which is sent to the *Cartesian robot controller* for actuation. Controller gains k_1 and k_2 in equations (6.6)-(6.7) are chosen to minimize overshoot and ringing due to measurement and actuation delays. The control cycle is repeated for each new frame until the pose error (equations (6.2)-(6.3)) is sufficiently small: $|{}^{G}\mathbf{T}_O| < d_{th}$ and ${}^{G}\theta_O < \theta_{th}$, where ${}^{G}\theta_O$ is from the axis/angle representation of the orientation error, and d_{th} and θ_{th} are the pose error thresholds.

6.6.1 Image Processing

In this implementation, image processing is deliberately simplified to facilitate robust, real-time visual servoing. Stereo images are captured at PAL frame-rate (25 Hz), 384×288 pixel resolution and 16 bit RGB colour. Image processing is implemented on a desktop PC, taking full advantage of the MMX/SSE instruction set and multiple processors to parallelize operations where possible (see Section 7.1.2). Also, image processing is restricted to a sub-image region of interest (ROI) most likely to contain valid measurements. The ROI is constructed as the bounding box enclosing the marker locations projected onto the image planes (equation (6.11)) at the predicted pose of the end-effector (equation (6.31)). The main image processing task during servoing is to measure the position of each active marker, implemented as a red LED, while robustly handling the possibility of occlusions and background clutter. LED detection is implemented as a two step process: colour filtering identifies a set of candidate LED locations, and candidates are then matched to the predicted LED positions using a global matching algorithm.

The colour filtering step identifies pixels similar in colour to the LEDs. The filter is implemented as a look-up table (LUT) with a binary entry for each 16-bit RGB value indicating membership in the pass-band. The LUT is constructed from visual measurements in the following once-off calibration process. While viewing the stationary gripper, the LEDs are alternately lit and darkened in successive captured frames, so that the inter-frame pixel difference identifies foreground pixels on the LEDs. The foreground colours are accumulated over several frames in a colour histogram (one cell for each 16-bit RGB value). Finally, histogram cells with less

(a) Gripper against background clutter

(b) Output of colour/morphological filter

Fig. 6.4. Detection of LED candidates via colour filtering.

than 20% of the maximum value are cleared, and the remaining non-empty cells are thresholded to form the colour pass-band. The filter is applied to captured frames by replacing each pixel with the corresponding entry in the LUT. The resulting binary image is morphologically eroded and dilated to reduce noise, and binary connectivity analysis is applied to identify connected blobs. The centre of mass of pixels in each blob serve as the initial candidate LED locations.

Figure 6.4(a) shows a typical view of the gripper against a cluttered background. The output of the colour/morphological filter for the same scene is shown in Figure 6.4(b); a number of LEDs remain undetected while background clutter introduces spurious features. A global matching algorithm helps to identify the visible LEDs while rejecting noise. The first step in this algorithm is to predict the LED locations by projecting the gripper model in the predicted pose. Let $\hat{\mathbf{g}}_i$, $i = 1 \dots n$, represent the predicted position of the n LEDs from equation (6.11), and $\mathbf{g}_j$, $j = 1 \dots m$, represent the m candidate measurements (noting that m may be larger or smaller than n). Now, denoting the association of LED i with candidate measurement j as a_{ij}, the aim of the matching algorithm is to find the most likely set of associations $A = \{a_{ij}\}$ in which each of i and j appears zero or one times, ie. $|A| \leq \min(m,n)$.

A simple matching approach is to associate each LED with the closest candidate measurement. However, this only works when the predicted pose is reliable and few spurious features are present. At the other extreme, a brute force matching search evaluates the likelihood of *all* possible associations, which can be very costly for large m. The algorithm described below avoids both problems using a sub-optimal global matching process that is more reliable than closest-point matching, but requires fewer computations than a brute force search.

Global matching aims to identify LEDs and candidates that are similarly placed relative to surrounding matched features, as illustrated in Figure 6.5. Let $\hat{\mathbf{d}}_{ik} = \hat{\mathbf{g}}_k - \hat{\mathbf{g}}_i$ denote the position of the kth predicted feature relative to the ith predicted feature, and $\mathbf{d}_{jl} = \mathbf{g}_l - \mathbf{g}_j$ denote an equivalent relative measure between candidates. To evaluate association a_{ij} in Figure 6.5, the algorithm searches for other LED/candidate

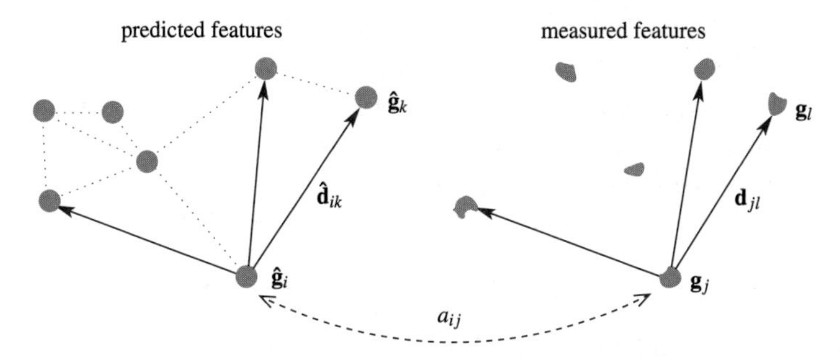

Fig. 6.5. Solving the association between predicted and measured features. The association a_{ij} between the predicted LED $\hat{\mathbf{g}}_i$ and candidate measurement $\mathbf{g}_j$ is supported by other prediction/measurement pairs with matching relative positions. In this example, three prediction/measurement pairs (including $\hat{\mathbf{g}}_k$ and $\mathbf{g}_l$) are found to support association a_{ij}. *Reprinted from [149]. ©2004, IEEE. Used with permission.*

pairs with similar positions relative to $\hat{\mathbf{g}}_i$ and $\mathbf{g}_j$ (three such pairs are shown in Figure 6.5). The number n_{ij} of pairs consistent with the association a_{ij} is calculated as

$$n_{ij} = \sum_k \begin{cases} 1 \text{ if } \min_l |\hat{\mathbf{d}}_{ik} - \mathbf{d}_{jl}| < \varepsilon_{th} \\ 0 \text{ otherwise} \end{cases} \tag{6.37}$$

The threshold ε_{th} is the maximum allowed difference in relative position for a good match. A residual error r_{ij} is calculated for association a_{ij} as

$$r_{ij} = \frac{1}{n_{ij}} \sum_k \begin{cases} \min_l |\hat{\mathbf{d}}_{ik} - \mathbf{d}_{jl}| & \text{if } \min_l |\hat{\mathbf{d}}_{ik} - \mathbf{d}_{jl}| < \varepsilon_{th} \\ 0 & \text{otherwise} \end{cases} \tag{6.38}$$

The above formulation always gives $n_{ij} \geq 1$, since the LED/candidate pair given by $k = i$ and $l = j$ always matches the association a_{ij}. Treating n_{ik} as the likelihood of the association a_{ik}, the best match for LED i is the candidate j with the maximum score, $n_{ij} = \max_k n_{ik}$. The best association for each LED is collected in the set L (note that LED i does not appear in L if $n_{ij} \leq 1$):

$$L = \{a_{ij} : n_{ij} = \max_k n_{ik},\ n_{ij} > 1\} \tag{6.39}$$

If the maximum score is shared by more than one association for a given LED, the ambiguity is resolved by selecting the association with the minimum residual error, $r_{ij} = \min_k r_{ik}$. The association is then performed in reverse, finding the LED i that best matches candidate j over all LEDs. The reverse associations are stored in set C:

$$C = \{a_{ij} : n_{ij} = \max_k n_{kj},\ n_{ij} > 1\} \tag{6.40}$$

The set A' of mutually supporting associations a_{ij}, where candidate j is the best match for LED i and *vice versa*, is calculated as the intersection

$$A' = L \cap C \tag{6.41}$$

Next, the associations in A' are tested for self-consistency. Consider the associations a_{ij} and a_{pq}: if consistent, the displacement between LEDs i and p should be similar to the displacement between candidates j and q. Using this rule, the set A_{ij} of associations $a_{pq} \in A'$ consistent with $a_{ij} \in A'$ is calculated as

$$A_{ij} = \{a_{pq} : |\hat{\mathbf{d}}_{ip} - \mathbf{d}_{jq}| < \varepsilon_{th},\ a_{pq} \in A'\} \tag{6.42}$$

Finally, the largest self-consistent set A, given by

$$A = \operatorname{argmax}|A_{ij}| \tag{6.43}$$

is taken as giving the best associations between measurements and LEDs. LEDs not appearing in A are assumed to be obscured, and successfully matched measurements are passed to the measurement vector of the IEKF.

6.6.2 Initialization and Calibration

The IEKF solves the non-linear system equations numerically, so a good estimate of the initial pose and robot position are required to initialize the filter at the start of a new servoing task. Typically, these parameters should be known from the output of the filter in previous tasks. When this information is unavailable, the autonomous calibration procedure described in this section is applied before servoing commences. This process exploits the fact that the gripper features are actively controlled to solve the measurement association problem without measurement prediction. Note, however, that active association cannot be used during servoing since it requires a static end-effector.

The calibration process begins by locating the gripper. With all LEDs illuminated, the cameras are scanned across the workspace to find the direction giving maximum output from the colour filter. The LEDs are then flashed in sequence and measured separately to avoid association errors. Colour filtering and image differencing are applied to successive frames to measure the location of individual LEDs. Finally, the initial pose ${}^W\hat{\mathbf{p}}_E(0)$ and scale factor $\widehat{K}_1(0)$ of the gripper are estimated by minimizing the reprojection error as described in Section 6.3 using the Levenberg-Marquardt (LM) algorithm (as implemented in MINPACK [105]). Being a numerical method, LM optimization also requires an initial estimate of the pose and scale. Reliable results are obtained by setting the scale to unity, orientation to zero, and estimating the translation from equation (6.14) using the average position of LEDs on each image plane.

Once the pose is known, the robot position is initialized as ${}^W\mathrm{H}_B = {}^W\mathrm{H}_E({}^B\mathrm{H}_E)^{-1}$, where ${}^B\mathrm{H}_E$ is estimated from the joint angles and kinematic model. Finally, the initial pose and robot position are passed to the IEKF and the state error variances are set to large values. This ensures that the estimated state is initially weighted towards new measurements until the state error covariance settles to a steady-state value.

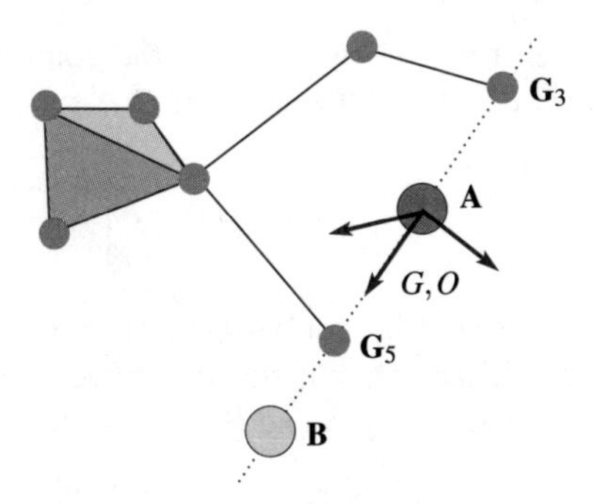

Fig. 6.6. Experimental task for evaluation of hybrid position-based visual servoing.

6.7 Experimental Results

This section presents several experimental trials designed to evaluate the accuracy and robustness of hybrid position-based visual servoing, and compare this approach to classical techniques. Further experiments with visual servoing in practical domestic tasks are presented in Chapter 7. All experiments in this chapter involve the positioning task is illustrated in Figure 6.6. The task involves a simplified target consisting of two coloured markers located at **A** and **B**. The goal is to position the thumb and index finger of the gripper ($\mathbf{G}_5$ and $\mathbf{G}_3$) on the line intersecting **A** and **B**, placing **A** at the midpoint between the fingers.

Section 6.2 defines the grasp frame G and object frame O, which are aligned at the completion of the servoing task. In this case, the origin of the grasp frame G is set to the midpoint between the thumb and index fingers, with the Y-axis oriented towards the thumb. In this configuration, the translational and rotational components of ${}^{E}\mathrm{H}_{G}$ are

$$ {}^{E}\mathbf{T}_{G} = \frac{1}{2}({}^{E}\mathbf{G}_5 + {}^{E}\mathbf{G}_3)^{\top} \tag{6.44} $$

$$ {}^{E}\mathrm{R}_{G}(\theta, \mathbf{a}) = \mathrm{R}\left(\cos^{-1}\left(\mathbf{Y}\cdot\frac{{}^{E}\mathbf{G}_5 - {}^{E}\mathbf{G}_3}{|{}^{E}\mathbf{G}_5 - {}^{E}\mathbf{G}_3|}\right), \mathbf{Y}\times\frac{{}^{E}\mathbf{G}_5 - {}^{E}\mathbf{G}_3}{|{}^{E}\mathbf{G}_5 - {}^{E}\mathbf{G}_3|}\right) \tag{6.45} $$

where $\mathrm{R}(\theta, \mathbf{a})$ is the rotation matrix for the given angle and axis, and $\mathbf{Y} = (0,1,0)^{\top}$ is a unit vector in the direction of the Y-axis. Similarly, the object frame O is centred at **A**, with the Y-axis pointing towards **B**. Frame O is oriented so that the fingers point to the right in the desired pose (since the robot is left-handed!) as shown in Figure 6.6. The translational and rotational components of ${}^{W}\mathrm{H}_{O}$ are

$$ {}^{W}\mathbf{T}_{O} = \widehat{\mathbf{A}} \tag{6.46} $$

$$ {}^{W}\mathrm{R}_{0}(\theta, \mathbf{a}) = \mathrm{R}\left(\cos^{-1}\left(\mathbf{Y}\cdot\frac{\widehat{\mathbf{B}} - \widehat{\mathbf{A}}}{|\widehat{\mathbf{B}} - \widehat{\mathbf{A}}|}\right), \mathbf{Y}\times\frac{\widehat{\mathbf{B}} - \widehat{\mathbf{A}}}{|\widehat{\mathbf{B}} - \widehat{\mathbf{A}}|}\right)\cdot\mathrm{R}(\pi/2, \mathbf{Y}) \tag{6.47} $$

where $\widehat{\mathbf{A}}$ and $\widehat{\mathbf{B}}$ are the measured locations of the markers, reconstructed from stereo measurements of the colour centroids (using equation (6.14)).

After completing each trial, the accuracy of the controller is measured as the translational error e_T between $\widehat{\mathbf{A}}$ and the origin of G:

$$e_T = \left| \frac{1}{2}(\widehat{\mathbf{G}}_3 + \widehat{\mathbf{G}}_5) - \widehat{\mathbf{A}} \right| \tag{6.48}$$

where $\widehat{\mathbf{G}}_3$ and $\widehat{\mathbf{G}}_5$ are recovered by applying equation (6.14) to the stereo colour centroids of the associated LEDs. The rotational accuracy is measured as the angular error e_θ between the line intersecting $\widehat{\mathbf{A}}$ and $\widehat{\mathbf{B}}$ and the line intersecting $\widehat{\mathbf{G}}_3$ and $\widehat{\mathbf{G}}_5$:

$$e_\theta = \cos^{-1}\left(\frac{\widehat{\mathbf{G}}_5 - \widehat{\mathbf{G}}_3}{|\widehat{\mathbf{G}}_5 - \widehat{\mathbf{G}}_3|} \cdot \frac{\widehat{\mathbf{B}} - \widehat{\mathbf{A}}}{|\widehat{\mathbf{B}} - \widehat{\mathbf{A}}|} \right) \tag{6.49}$$

The pose errors e_T and e_θ are averaged over a number of frames to remove uncertainties in the reconstructed features. Finally, it should be noted that while the errors in equations (6.48)-(6.49) are not metric, these measurements expose any bias in the IEKF which may affect the performance of the controller. The following section evaluates the accuracy of hybrid position-based visual servoing compared to conventional ECL and EOL control. Section 6.7.2 then evaluates the robustness of hybrid control in the presence of clutter and occlusions. Lastly, Section 6.7.3 tests the limits of camera calibration errors handled by the IEKF.

6.7.1 Positioning Accuracy

This series of experiments evaluates the accuracy of the hybrid position-based visual servoing and conventional ECL and EOL position-based schemes under favourable operating conditions (minimum visual clutter, accurate calibration and small initial pose error). The positioning task was repeated over five trials for each controller, from the initial pose shown in Figure 6.7(a). In all cases, the pose error thresholds for convergence were set to $d_{th} = 5$ mm for translation and $\theta_{th} = 0.2$ rad for orientation, which are suitable levels of accuracy for typical service tasks.

Evaluating the hybrid controller first, Figure 6.7(b) shows the view at the end of a typical trial. The estimated gripper is overlaid as a yellow wireframe, and coordinate frames representing O and G are overlaid in red and blue respectively. Figure 6.8 plots the pose error during servoing, which exhibits typical behaviour for a proportional velocity controller. Next, a conventional ECL controller was implemented by excluding kinematic measurements from the IEKF measurement vector, excluding ${}^B\mathrm{H}_W$ and K_1 from the state vector and setting the fixed value $K_1 = 1$. Figure 6.9 shows the final pose of the gripper for a typical trial with the ECL controller. Finally, an EOL controller was implemented by discarding the IEKF entirely and estimating the pose of the end-effector purely from kinematics measurements (including a manually calibrated ${}^W\mathrm{H}_B$). For positioning trials using the EOL controller, a typical final pose is shown in Figure 6.10.

The average error and error variance calculated using equations (6.48)-(6.49) and averaged over all trials for each controller are shown in Table 6.1. The low variance

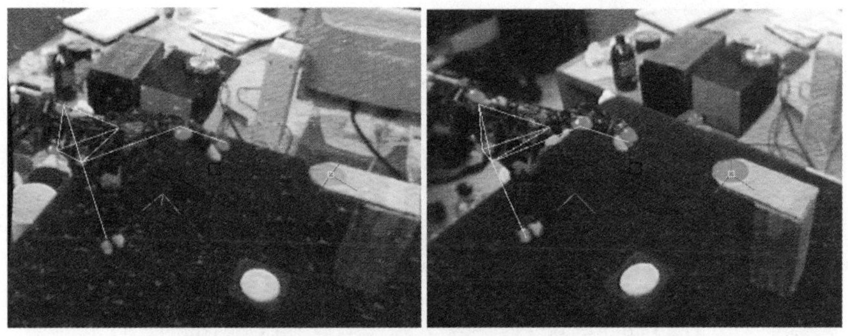

(a) Stereo images showing initial pose of object and gripper.

(b) Stereo images showing final pose at completion of servoing task.

Fig. 6.7. Positioning task result with hybrid position-based servoing. The estimated pose of the gripper is indicated by a yellow wireframe model, while the coordinate axes of the gripper and object frames are shown in blue and red respectively.

for all controllers reflects the high repeatability of the results and indicates that the residual pose errors e_T and e_θ are systematic rather than random. As expected, the EOL controller is significantly less accurate than the controllers using visual feedback, since the estimated pose is biased by calibration errors in the hand-eye transformation. The full ECL controller compensates for both camera and kinematic calibration errors and subsequently achieves the greatest accuracy. In fact, the final translation error is bounded only by the servoing termination threshold ($d_{th} = 5$ mm). The orientation accuracy is better than the termination threshold since orientation converges faster than translation for the chosen gains (see Figure 6.8). As expected, the ECL control achieves an accuracy somewhere between the other methods. The high accuracy achieved by hybrid control compared to ECL control is mainly due to the estimation of K_1 to compensate for camera calibration errors. This result highlights the importance of camera calibration for accurate position-based visual servoing, and demonstrates that this can be achieved on-line for mild errors by simply introducing a scale parameter associated with the estimated pose.

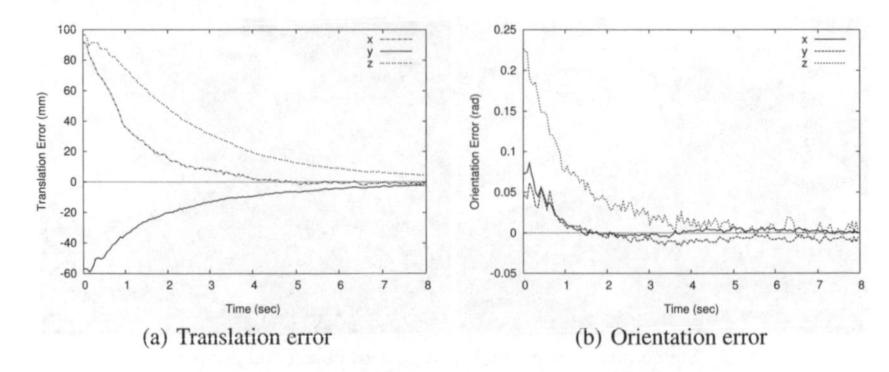

(a) Translation error (b) Orientation error

Fig. 6.8. Pose error for hybrid position-based visual servoing.

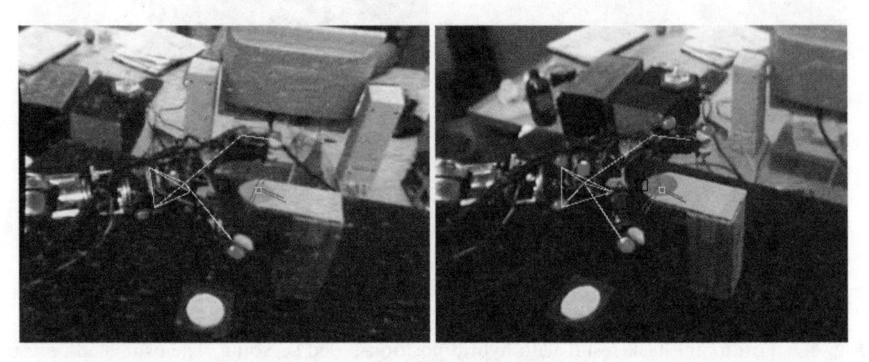

Fig. 6.9. Positioning task result with conventional ECL position-based controller.

Fig. 6.10. Positioning task result with conventional EOL position-based controller.

6.7.2 Tracking Robustness

This second experiment evaluates the robustness of hybrid position-based servoing in poor visual conditions including occlusions, background clutter and large initial

Table 6.1. Comparison of positioning accuracy for experimental visual servo controllers.

Controller	e_T (mm)	$var(e_T)$ (mm^2)	e_θ (rad)	$var(e_\theta)$ (rad^2)
Hybrid position-based	5.4	0.4	0.11	0.003
ECL: no scale correction	29.4	0.9	0.22	0.01
EOL: kinematics only	54.5	1.6	0.18	0.008

pose error. Figure 6.11(a) shows the initial pose of the object against a cluttered background. A large initial pose error places the end-effector outside the field of view. To test the effect of occlusions, a *virtual 3D obstacle* is rendered near the object on the left image plane.

After servoing commences and before the gripper enters the field of view, the pose is tracked using only kinematic feedback. Figure 6.11(b) shows the estimated pose (as a wireframe overlay) during visual servoing, just after the gripper becomes visible. At this point, the controller switches from EOL control to full hybrid control. As the gripper approaches the target, measurements on the left image plane are occluded by the virtual obstacle. At this point, estimation of K_1 is suspended since only monocular measurements are available. Finally, Figure 6.11(c) shows the successful completion of the task.

The performance of the controller is indicated by the plots in Figure 6.12, which show the translational pose error, standard deviation of the gripper translation reported by the IEKF and the estimated value of K_1. During the first three seconds, the gripper is outside the field of view and the scale remains at the initial value of $K_1 = 1$. The uncertainty in the end-effector pose during this period (see Figure 6.12(b)) is over-optimistic, since the pose is biased by the imperfectly calibrated hand-eye transformation. When stereo measurements finally become available, the estimated scale quickly converges to the actual value and the translation error is mildly perturbed as the filter adjusts to the new measurements. Furthermore, the additional constraints cause the pose uncertainty to decrease.

The gripper is obscured from view by the virtual obstacle approximately six seconds into the task. With only monocular measurements, K_1 is no longer sufficiently constrained and remains fixed at the most recent estimate. However, the hand-eye transformation (not shown) continues to be updated, based on the fixed scale. The controller achieves a final translation error of $e_T = 10.9$ mm and orientation error of $e_\theta = 0.087$ radian (the virtual obstacle is removed to complete these measurements), indicating only a mild reduction in accuracy from the case in Section 6.7.1 of stereo measurements and favourable visual conditions.

6.7.3 Effect of Camera Calibration Errors

The effect of calibration errors on the accuracy of visual servoing was examined in Section 6.3, and the experimental results in Section 6.7.1 verified that mild errors can be compensated by incorporating a scale factor K_1. This final experiment tests the bounds of the error model by observing the effect of deliberately introduced cal-

(a) Initial pose of object and virtual obstacle (shown in grey). The gripper is outside the field of view and initially tracked using only kinematic measurements.

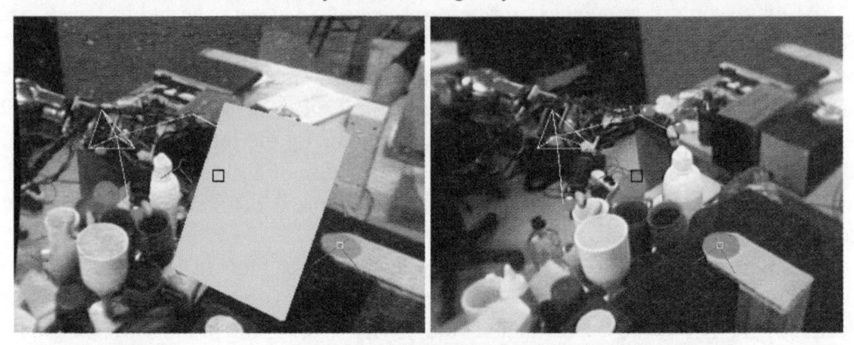

(b) Gripper enters the field of view and visual feedback is initiated.

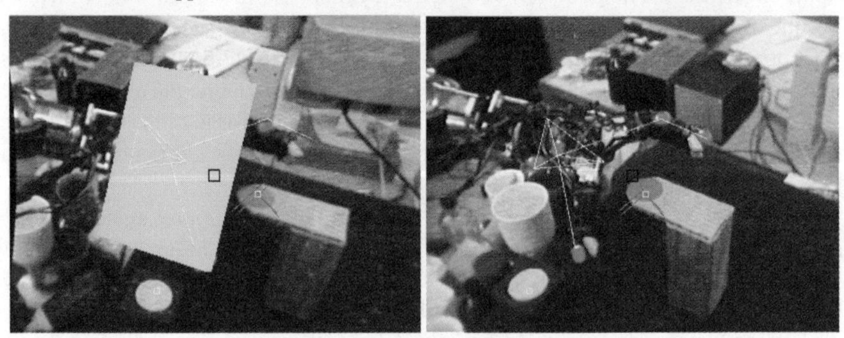

(c) Final pose at completion of the servoing task. The gripper is obscured by the virtual obstacle in the left visual field, leaving only monocular feedback.

Fig. 6.11. Select frames from positioning task in the presence of occlusions and clutter.

ibration errors. Specifically, the calibrated baseline (*not* the mechanical baseline) is scaled by 0.7 to 1.5 times the nominal value, and the verge angle is offset by -0.07 to 0.14 radian (-4 to +8 degrees) from the nominal measured value. The positioning task was performed using the hybrid position-based controller with the initial con-

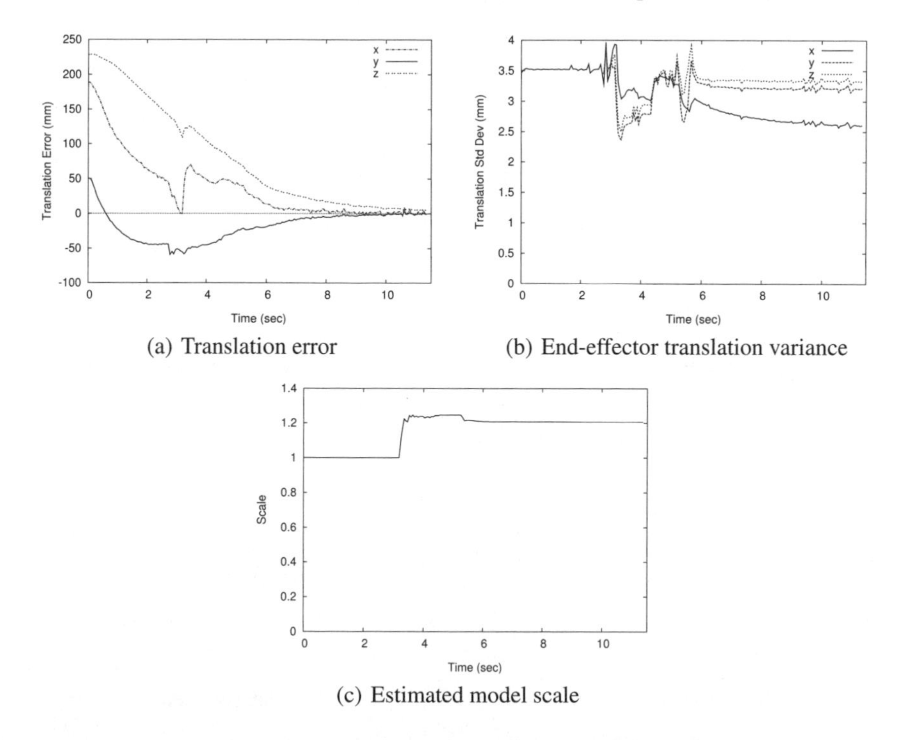

(a) Translation error

(b) End-effector translation variance

(c) Estimated model scale

Fig. 6.12. Visual servoing performance in the presence of visual occlusions and clutter.

figuration shown in Figure 6.7(a). At the completion of each trial, the final estimated scale was recorded along with the pose error.

Figure 6.13 shows the final estimated scale and servoing accuracy for trials with varying error in the baseline. The results support the expected linear relationship between baseline error and estimated K_1, as predicted by equation (6.20). Furthermore, Figures 6.13(b) and 6.13(c) show that the positioning accuracy of the hybrid is not significantly affected by the baseline error. This is also expected, since the scale-only error model is exact for baseline errors (that is, b is a factor of $\widehat{\mathbf{X}}$ in equation (6.14)). However, it should be noted that the constant position error reported in Figure 6.13(b) actually corresponds to a linearly increasing real-space error, due to the variation in the baseline (that is, the pose error is measured in a non-metric space).

The final scale and pose error for trials with varying verge angle error are shown in Figure 6.14. The main limitations of the verge angle error model arise from the small angle approximations used to derive equation (6.17). As the verge angle increasingly deviates from the calibrated value, non-linearities bias the estimated pose and decrease the accuracy of the controller. This trend is reflected in the observed pose errors in Figures 6.14(b) and 6.14(c), which exhibit a local minima. It should be noted that the minimum pose error should occur at the offset for which the estimated scale is unity (at a verge offset of 0.04 rad from Figure 6.14(a)), which is

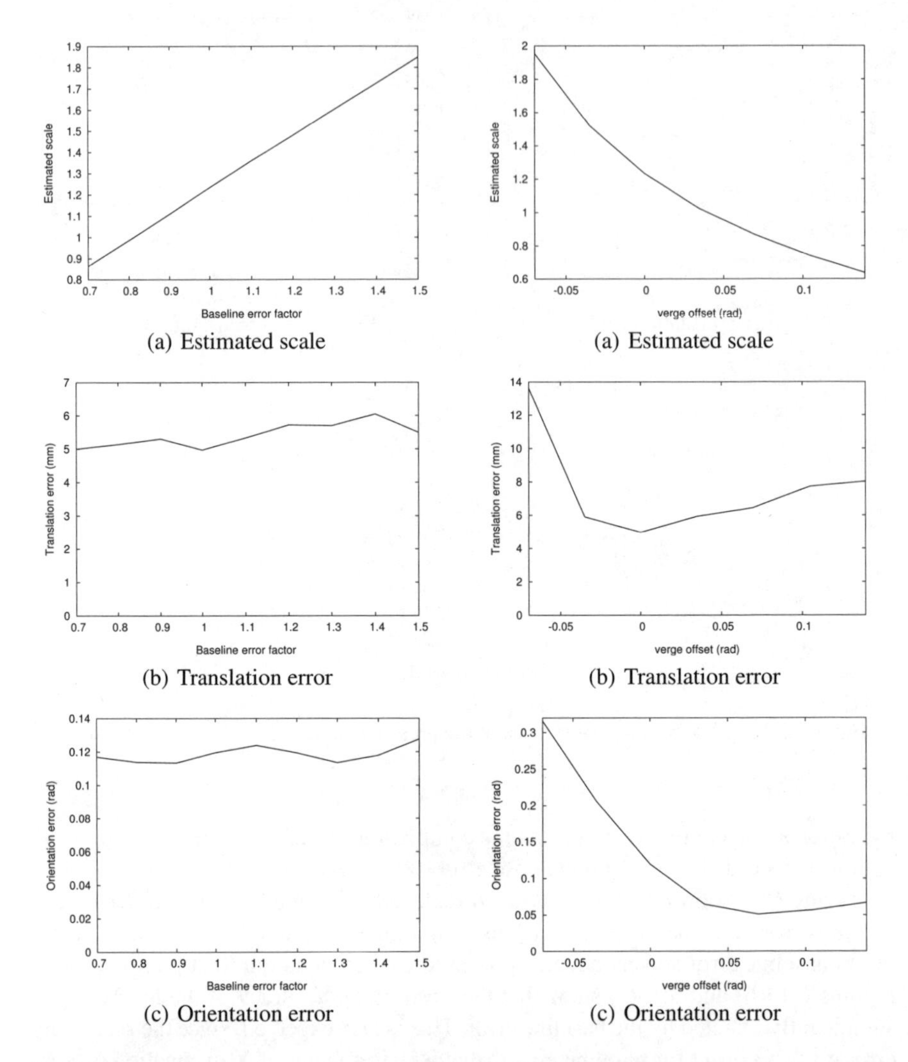

(a) Estimated scale (b) Translation error (c) Orientation error

Fig. 6.13. Performance of proposed controller for baseline calibration error.

(a) Estimated scale (b) Translation error (c) Orientation error

Fig. 6.14. Performance of proposed controller for verge calibration error.

clearly not the case. This discrepancy is most likely explained by the presence of an unknown baseline error in addition to the verge offset.

6.8 Discussion and Conclusions

This chapter presented the development and implementation of a hybrid position-based visual servoing framework suitable for service tasks. The proposed framework places particular emphasis on robustness to the loss of visual feedback, and on-line compensation for calibration errors, which are important factors in an uncontrolled environment. Conventional position-based visual servoing schemes employ either kinematic (EOL) or visual (ECL) feedback, and are therefore susceptible to kinematic calibration errors and visual occlusions. The robust end-effector pose estimator proposed in this chapter was designed to avoid both issues by optimally fusing both kinematic and visual measurements. Kinematic measurements allow the controller to operate (with reduced accuracy) while the gripper is occluded, and visual measurements provide accurate pose control and on-line estimation of kinematic parameters. Experimental results verify both the increased accuracy gained with visual feedback and the robustness to occlusions gained with kinematic feedback.

Sensitivity to camera calibration errors is usually considered the primary drawback of position-based visual servoing, and it was shown in Section 6.3 that a visually estimated pose is indeed biased by camera calibration errors. It was then shown that the effect of mild baseline and verge angle errors can be modelled as an unknown scale. While the error model considers only baseline and verge angle errors, this is not an unreasonable limitation for many robotic implementations. The baseline and verge angle both vary as a pan/tilt head is actuated, while other camera parameters are fixed and typically well known or readily estimated from visual measurements using a variety of established calibration techniques. Experimental results validated the error model and demonstrated that the accuracy of the controller is improved by estimating the unknown scale along with the pose of the end-effector. The combination of online calibration and kinematic/visual fusion described in this chapter overcomes many of the classical problems associated with position-based visual servoing.

The error model developed in this chapter could also be applied to the multi-cue 3D model-based tracker developed in Chapter 5. However, new considerations arise for tracking multiple objects (for example, the gripper and target object) in the same image. Each object is associated with a different scale error, since the scale depends on the current pose (through K_1 in equation (6.20)). In addition, any change in scale due to camera motion (in particular, the verge angle) will be correlated for all objects. Tracking should account for this correlation to optimally estimate the scale of each object. In practice, the objects are modelled from reconstructed 3D measurements that are already biased by the scale error (unlike the gripper, which is calibrated from metric measurements). Furthermore, the verge angle does not vary significantly during servoing after fixating on the target. As a result, object tracking using data-driven models is not significantly affected by calibration errors.

Lastly, it should be mentioned that the controller presented in Section 6.2 overlooks dynamic control issues such as steady state error (for moving targets) and sensing/actuation delays. In the experimental work, these issues were avoided by using sufficiently low gain and stationary targets. The system actually suffers from an acquisition delay of at least 40 ms between capturing and processing stereo images, due to internal buffering. Furthermore, an actuation delay of about 170 ms has been observed between issuing a velocity command to the Puma robot arm and observing the subsequent motion. Appendix E describes the simple calibration procedure that was used to measure this and other latencies in the system. Clearly, dynamic control issues must be addressed before a service robot can perform with the speed and flexibility of a human. A detailed treatment of dynamic control issues in visual servoing can be found in [27].

7

System Integration and Experimental Results

In this chapter, a hand-eye service robot capable of performing interactive tasks in a domestic environment is finally realized. The shape recovery, object modelling and classification, tracking and visual servoing components developed in earlier chapters are integrated into a framework illustrated by the block diagram in Figure 1.2 (page 7). For the experiments in this book, the system is implemented on an upper-torso humanoid robot. Humanoid robots are an appropriate test-bed since hand-eye co-ordination will be the next vital skill once humanoid robot research moves beyond bipedal motion. But just as importantly, a humanoid robot engenders a much greater sense of personality than other configurations! It is important to note that the methods in this book also apply to a range of camera and manipulator configurations, anthropomorphic or otherwise. The experimental platform, known as Metalman, is detailed in Section 7.1.

The three fundamental steps in performing any task are: specification, planning and execution. Specification is the problem of supplying the robot with sufficient information to perform the required task, in a form that is unambiguous and easy to interpret for both the user and robot. Section 7.2 outlines the common approaches to task specification and details the methods adopted in this chapter. The planning component then uses the specifications to generate actions aimed at achieving the desired goal. For this step, Section 7.3 describes a algorithm to plan a collision-free path for the end-effector to grasp an object. Further details of the task planning calculations are also presented in Appendix F. Finally, the planned path is passed to the visual servo controller to execute the manipulation. Unlike industrial robots, service robots may require several iterations between specification, planning and execution as unforeseen or ambiguous obstacles are encountered while performing a task.

Three domestic tasks are implemented to evaluate the performance of the proposed framework. In the first experiment, the robot is required to locate and grasp a yellow box with minimal prior knowledge of the scene. The target object is only specified in terms of shape class (box) and colour (yellow), and is placed amongst several other objects on a table. The second experiment requires the robot to pick up a cup that is interactively selected by the user, and pour the contents into a bowl. Again, the robot has no prior knowledge of the scene except the class of the targets.

The third task explores the possibility of augmenting visual sensing with airflow and odour sensors to increase the utility of the robot. This is demonstrated by performing a recognition task that cannot be accomplished by vision alone; the robot is required to autonomously locate and grasp a cup containing ethanol from amongst several objects (including other cups).

The implementation and experimental results for the three task are presented in Sections 7.4, 7.5 and 7.6, and additional VRML data and video clips can be found in the *Multimedia Extensions*. It is important to note that a practical service robot will require many additional skills, including dextrous manipulation, compliant joint control, and high-level human interaction, that are beyond the scope of this book. By necessity, the tasks considered in this chapter are also limited by the capabilities of the experimental platform, and are appropriately contrived to avoid sophisticated task planning. Despite these limitations, the experimental results presented in this chapter demonstrate the significant steps that this work has taken towards the development of useful and general hand-eye robotic systems.

7.1 Experimental Hand-Eye Robot Platform

The following sections describe the hardware and software architecture of the experimental upper-torso humanoid platform used in this work.

7.1.1 Hardware Configuration

The experimental robotic platform, known as Metalman, is shown in Figure 7.1, and a block diagram of the major components is illustrated in Figure 7.2. Metalman is an upper-torso humanoid robot, with most of the control hardware located off-board. The robot is approximately anthropomorphic in scale and configuration, with six degree of freedom (DOF) arms and stereo cameras on a three DOF active head. While the lack of degrees of freedom about the waist limits the viewpoint and workspace of the robot, the available motion is sufficient for the study of manipulation in unstructured environments. It should also be noted that only one of the arms was used for the tasks implemented in this chapter, leaving two arm coordination for future work. While many humanoid robot platforms employ purpose-built hardware, Metalman is mainly constructed from off-the-shelf components to aid simplicity and functionality. The following sections detail the major components of the system:

Processing Platform

The processing platform is a dual 2.2GHz Intel Xeon PC operating Red Hat Linux 7.3. The two Xeon processors are each capable of simultaneously processing two independent instruction streams, and the system is therefore capable of executing up to four concurrent processes. Furthermore, integer and floating point SIMD operations in the Xeon allow up to 16 pixels to be processed in parallel. By exploiting both features (see section 7.1.2), all signal processing and high-level hardware control is implemented on board the PC.

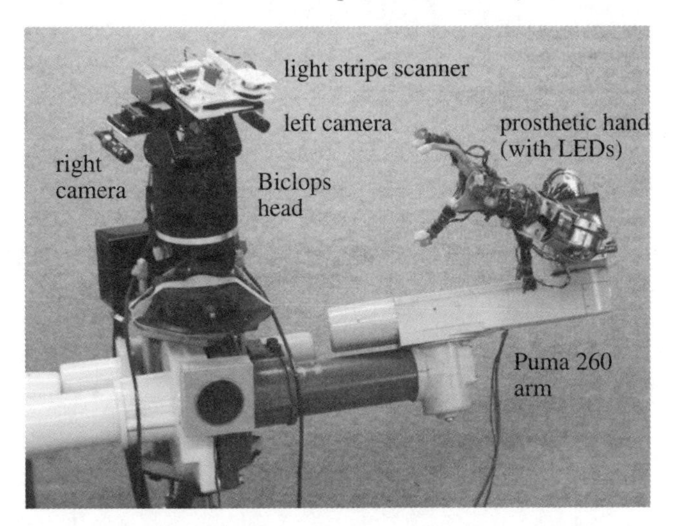

Fig. 7.1. Metalman: an experimental upper-torso humanoid robot. *Reprinted from [150]. ©2004, IEEE. Used with permission.*

Stereo Cameras

Vision is provided by a pair of Panasonic CD1 CCD camera systems, which capture images at PAL frame-rate (25 Hz) and half-PAL (384×288 pixel) resolution. The Genlock mechanism is used to ensure the simultaneous exposure of stereo fields, which is important when capturing dynamic scenes. Images are digitized in the host PC by analogue video capture cards based on the Brooktree BT878 chipset.

Biclops Head

The stereo cameras are mounted on a three DOF Biclops head. In addition to the usual pan and tilt axes, Biclops provides a non-linear verge control (see Appendix A). The orientation of the head is recovered via optical encoder measurements on each axis. Position and velocity control are implemented on a built-in PIC microcontroller, which communicates with the host PC via an RS-232 link.

3D Scanner

A stereoscopic laser stripe scanner mounted on the Biclops head provides dense range images of the workspace. The hardware consists of a 5 mW red laser (class IIIa) with a cylindrical lens to generate a vertical stripe, a DC motor/optical encoder and a PIC-based microprocessor to control the pan angle. The operation of the scanner is detailed in Chapter 3.

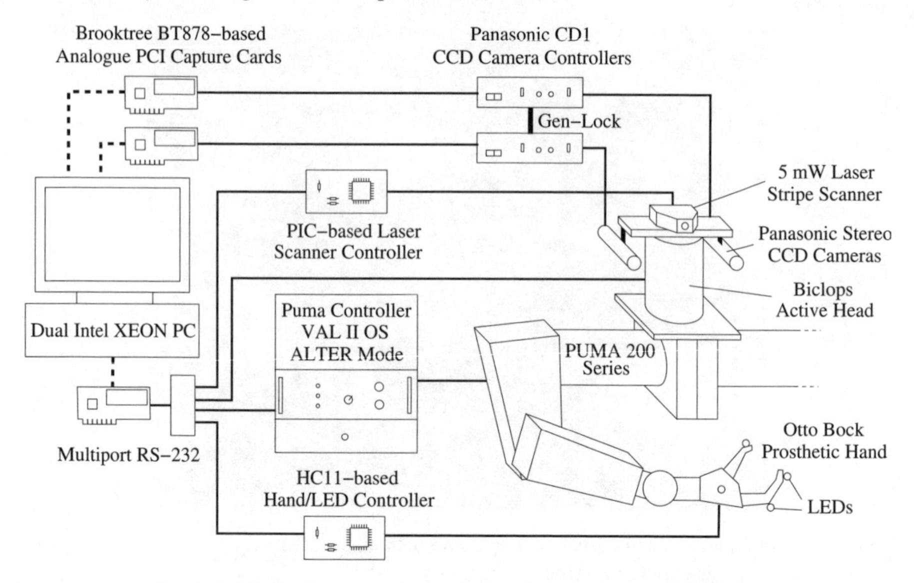

Fig. 7.2. Block diagram of components in experimental platform.

Puma Robot Arms

The two PUMA 200 Series robots serving as Metalman's arms are approximately anthropomorphic in configuration and scale, and are operated from the original LSI-11 control cabinets running VAL II. The PUMA controller communicates with the host PC over an RS-232 link in ALTER mode, accepting position/orientation corrections to a static nominal pose at regular 30 ms intervals (approximately matching the 40 ms sensing cycle of the PAL cameras). Pose corrections are specified in the tool-tip coordinate frame, and inverse kinematics and velocity profiling are provided by the PUMA controller.

Prosthetic Hands

Appended to the tool-tip of each PUMA arm is a single-DOF Otto Bock prosthetic hand with a parallel-jaw configuration. The limited range of manipulations afforded by this configuration is sufficient for simple pick-and-place tasks. The hands contain no tactile feedback, leaving Metalman to rely on the measured opening angle and visual feedback to determine the success of a grasp. Several LEDs are attached to the hand to serve as active features for visual tracking (see Chapter 6). The hands and LEDs are controlled by Motorola HC11 microcontrollers, which communicate with the host PC over RS-232 links.

7.1.2 Software Architecture

The software architecture emphasizes robustness and real-time operation rather than considerations of modularity and portability. In some cases, real-time image processing has been achieved through design, such as employing coloured LEDs for tracking the gripper and using prediction to identify a restricted *region of interest* for image processing. For the remaining processes, acceleration was achieved by parallelizing the computations where possible. The following sections describe the software tools used for this purpose.

Hardware Accelerated Processing: MMX/SSE/SSE2

MMX (multimedia extensions), SSE (streaming SIMD extensions) and SSE2 are single instruction, multiple data (SIMD) instruction sets for Intel processors that provide similar functionality to digital signal processing hardware [71]. The integer SIMD extensions provide 128-bit wide registers that can perform parallel operations on 1, 2, 4 or 8-byte integer operands, which is ideal for parallelizing many image processing operations. All of the pixel-point operations (threshold, subtraction, colour conversion, etc) and convolution operations (morphological filtering, edge extraction, template matching, etc) used in this research were parallelized using assembly coded SIMD to achieve a 4-8 times reduction in execution time compared to non-parallel implementations. However, not all image processing functions could be parallelized, most prominently the look-up table operations (such as colour filtering), and algorithms involving dependencies between adjacent pixels (such as connectivity analysis). SSE/SSE2 also support 4 packed single-precision or 2 double-precision floating point operands. Single-precision SIMD instructions were used to accelerate image processing functions such as radial correction/projective rectification and eigenvalue computations (see Chapter 5), while double-precision instructions were used to accelerate matrix operations in the Kalman filter.

Parallelization Using POSIX Threads

As noted earlier, each Xeon processor supports two *logical* processor states and increases pipeline efficiency by allowing instructions from one logical processor to occupy the wait states of the other. On the dual-Xeon platform, the hardware is presented to the operating system as four independent processors, and parallelization can therefore be exploited by dividing software into a number of concurrent processes (or *threads*). Thread management and inter-process communication was implemented using the POSIX threads model [17]. Image processing for each camera was handled by a separate thread, allowing the stereo fields to be processed concurrently. High-level results were periodically collected by the main program thread, which handled task planning and hardware control. Fine grained parallelization was also employed for some processor intensive operations. For example, calculation of the surface type for 3D range data (see Chapter 4) was accelerated by dividing the range image into four regions and processing each in a separate concurrent thread.

The main drawback of POSIX threads in the 2.4 Linux kernel is the inability to allocate threads to processors (known as assigning *processor affinity*)[1]. Dividing a program into threads did not guarantee a uniform distribution of load across all available processors, and the benefit of parallelization was often temporarily lost!

7.2 Task Specification

The purpose of task specification is to supply the robot with information regarding target objects and manipulations necessary to plan and execute the task. The most basic approach to task specification is to program a fixed manoeuvre either in software or by manually guiding the robot through a series of set-points. This approach is sufficient for the repetitive tasks that industrial robots are usually required to perform. Conversely, domestic tasks are characterized by arbitrary and often unforeseen manipulations involving possibly unknown objects in a dynamically changing environment, which precludes the use of pre-programmed manoeuvres. Furthermore, it cannot be assumed that the average user possesses sufficient technical expertise (or patience!) to program a new manoeuvre into the robot for every unique task.

The most promising solution to the task specification problem is the development of human-machine interfaces using natural modes of communication. Previous work on robotic task specification using human modes of interaction include verbal communication [131, 146], pointing gestures [24, 76], simple sign language [10, 66, 107, 122], and teaching by demonstration [32, 70]. Feedback from the robot to the user is also an important element of human-machine interaction. When an ambiguous task specification is encountered (for example, the requested object cannot be uniquely identified), the robot may seek clarification by requesting the user to choose between possible interpretations, as in [146].

The complexity of natural interaction techniques render them beyond the scope of this book. For simplicity, the experimental tasks presented in this chapter are pre-programmed as a scripted series of sub-tasks. Each sub-task describes an atomic operation such as locating a target, grasping, lifting, or aligning the grasped object with another object. The low-level set-points required for each manipulation sub-task are planned autonomously by the robot depending on environmental conditions (see Section 7.3). Object models are not provided directly, but are instead specified by class (cup, box, bowl, etc.), general features (colour, size, etc) or selected interactively by the user through a graphical user interface (GUI). While simplifying the implementation, this framework also demonstrates the basic elements of interaction and flexibility to deal with unknown objects and adapt to environmental conditions that are required by a useful service robot.

[1] Processor affinity is now supported in version 2.6 and above of the Linux kernel.

7.3 Task Planning

In this work, task planning is the process of determining the desired pose of the robot to grasp or place an object and generating a series of set-points to guide the gripper on a collision-free path towards the goal. The two stages in the process are described as *grasp planning* and *trajectory planning*.

For reach-and-grasp manipulations, a grasp planning algorithm searches for an optimal hand pose that results in stable contacts between the fingers and object. More sophisticated planners may also constrain the planned pose to avoid collisions with surrounding obstacles. Alternatively, the grasp planner may determine that the object cannot be grasped. A grasp is described as stable if the object does not slide or rotate about the contact points when lifted. For dextrous grippers and general objects, grasp planning is complex and computationally expensive, and has no closed form solution [38, 87]. Fortunately, grasp planning for Metalman is simplified by the single DOF parallel-jaw structure of the prosthetic hands. The process is simplified even further by developing a specific planner for each type of geometric primitive (box, cylinder, *etc.*), rather than a general solution for arbitrary shapes. This approach can be applied to any object modelled as a composite of primitives (see Chapter 4) by applying the appropriate planner to a suitable primitive component. Finally, this implementation does not consider collisions with surrounding obstacles when planning a grasp.

Referring to the coordinate frames defined in Figure 6.1 (page 119), the purpose of grasp planning is to determine the desired pose of the object (the grasp frame G) in the frame of the end-effector, represented by the transformation ${}^{E}\mathrm{H}_{G}$. The planners use either a *precision* or *power* grasp depending on the geometry of the component. A power grasp provides the greatest stability since the surface is contacted at multiple points along the fingers. However, a precision grasp (which makes contact only at the fingertips) is sometimes necessary when the fingers cannot wrap entirely around the object[2]. A power or precision grasp is generated by placing the origin of the grasp frame at the appropriate location within the plane of the fingers, represented by $\mathbf{G}_{pw}$ and $\mathbf{G}_{pr}$ and shown in Figure 7.3. Details of the grasp planning calculations to generate ${}^{E}\mathrm{H}_{G}$ for an arbitrary box and cup can be found in Appendix F, and are based on the grasp stability principles established in [37, 140].

Finally, trajectory planning generates a series of set-points to guide the end-effector from an arbitrary initial pose to the planned grasp. For high DOF redundant mechanisms such as a humanoid robot, trajectory planning in the presence of obstacles is a complex problem. Kuffner and LaValle [92] recently presented a promising solution based on *Rapidly-exploring Random Trees*. This algorithm searches through high dimensional configuration space using a process of random diffusion to plan manipulations in complex environments. To avoid these complexities, the experiments in this chapter are contrived to eliminate the problem of obstacle avoidance. The simplified task planner generates only a single intermediate set-point between the initial pose and planned grasp, as shown in Figure 7.4. This set-point ensures

[2] In general, a precision grasp facilitates dextrous manipulation, while a power grasp provides greater stability.

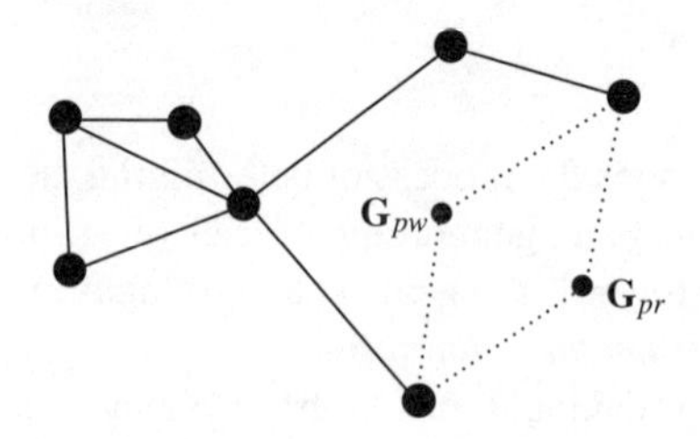

Fig. 7.3. Placement of grasp frame for power ($\mathbf{G}_{pw}$) and precision ($\mathbf{G}_{pr}$) grasps.

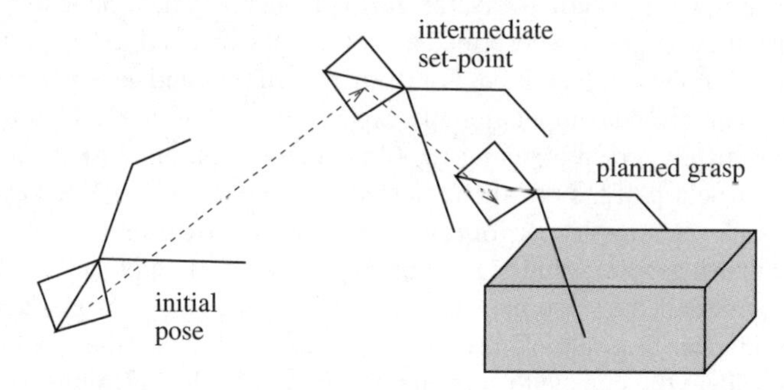

Fig. 7.4. Simplified grasp and trajectory planning for a box.

the fingers do not collide with the object as the gripper approaches the final planned grasp. Details of the intermediate set-point calculations are also provided in Appendix F.

7.4 Experiment 1: Grasping an Unknown Object

In this first experiment, Metalman is given the task of retrieving a specific object from a cluttered scene. However, the object is only specified by a generic class (a box) and particular feature (yellow colour). This task could have been conveyed to the robot using the verbal command: "Metalman, please fetch the yellow box". No other information about the scene (such as the number and type of objects) is known, and the end-effector is initially outside the field of view. The initial pose of Metalman and the arrangement of objects on the table are shown in Figure 7.5, with the target object farthest to the left of the robot. It will be assumed that the Puma base frame location and light stripe scanner have already been actively calibrated as described in Chapters 6 and 3. To complete this task, Metalman must identify an object satisfying the specifications and perform a grasp based on visual measurements of both the object and end-effector.

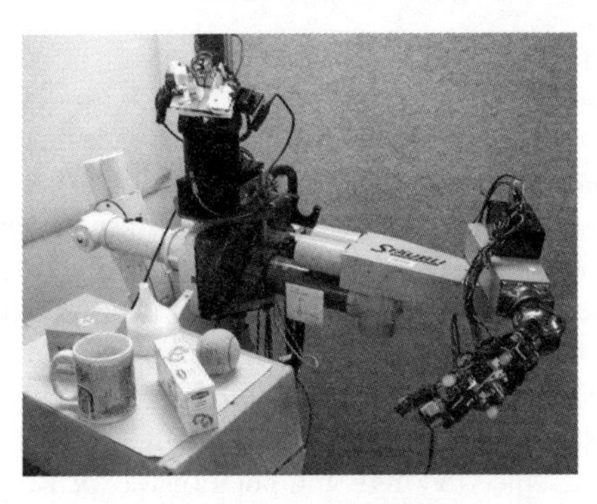

Fig. 7.5. Arrangement of objects for Experiment 1. Metalman is required to fetch the yellow box located farthest to the left of the robot. *Reprinted from [150]. ©2004, IEEE. Used with permission.*

7.4.1 Implementation

While object retrieval is a relatively straightforward operation, all of control and perception components developed in this book must be employed to achieve this goal. The steps to complete the task are described below:

Initialization

In the current implementation, the initial gaze of the robot is manually directed towards the table by interactively driving the Biclops head. For practical domestic robots, a more useful approach would combine visual perception with voice commands and a map of the local environment, as in the example: "Metalman, please fetch the yellow box from the kitchen bench". After manually establishing the initial pose, the remainder of the task is entirely autonomous.

Target Identification

To identify the specified object, the stereoscopic light stripe scanner first captures a 3D colour/range map of the scene as described in Chapter 3. The stripe is scanned in both the forward and reverse directions to ensure a good coverage of range measurements. The colour/range map is then segmented using the object classification algorithm in Chapter 4. The result is a list of convex objects and associated 3D polygonal models, with each object classified as either *box*, *ball*, *cup* or *bowl*. Metalman is required to select the object that best matches the description of a "yellow box", which is achieved by consulting the texture information for all detected boxes. The number of texture pixels (texels) within manually predefined ranges of hue, saturation and intensity for yellow are tallied, and the box with the highest number is selected as the desired target.

Grasp Planning and Execution

Using the planning processes described in Section 7.3, a final grasp and intermediate set-point are calculated for the target box. Finally, these are passed to the hybrid visual servo controller described in Chapter 6. Since the end-effector is initially outside the field of view, servoing commences using only kinematic measurements (EOL control). As the hand nears the object, visual measurements become available and the controller makes a transition to fused kinematic and visual measurements. To close the visual feedback loop, the pose of the object is continuously updated using the multi-cue tracking algorithm described in Chapter 5. The target is tracked at a reduced rate of 2 Hz, since the high computational expense of multi-cue tracking degrades the performance of the visual servoing at higher sample rates. Fortunately, a reduced rate is sufficient for static objects since tracking is primarily employed to overcome the pose bias due to calibration errors in ego-motion. Conversely, the moving gripper is tracked at full frame-rate.

The task is completed when the gripper reaches the planned pose and the object is grasped and lifted from the table. In the absence of tactile sensors, the joint angle of the fingers is monitored as the gripper is closed to determine when the object is stably grasped. The stability of the grasp could also be evaluated by visually tracking the object as it is lifted to detect slippage, although this technique is not used in the current implementation.

7.4.2 Results

Figure 7.6 shows two views of the unprocessed light stripe during the initial acquisition of range/colour measurements. Although this scene appears to pose significantly less challenge than the mirrored surface considered in Section 3.5.1, the circled regions of Figure 7.6 demonstrate how strong reflections can still arise from smooth cardboard and plastic to interfere with range data acquisition. Despite these distractions, the stereoscopic scanner produces a dense, accurate colour/range map as shown in Figure 7.7(a). Figure 7.7(b) shows the result of segmenting the range map into geometric primitives. The wireframe models for detected convex objects are overlaid on a captured frame in Figure 7.7(c), and the textured models are rendered together in Figure 7.7(d). VRML models of the raw colour/range data and extracted objects can be found in the *Multimedia Extensions*. Range data acquisition and segmentation are completed in a total time of about one minute.

Segmentation identifies two possible boxes: an orange box at the top right and a yellow box at the left, the latter being the desired target. The texture information from the extracted models is analysed to identify the yellow box as shown by the colour charts in Figure 7.8. The hue and saturation (intensity not shown) of texels are plotted as black points, and the predefined colour range for bright yellow is indicated by the black rectangle. The colour of the two boxes is clearly distinguished by the number of texels within the yellow range, and the desired box is correctly and unambiguously identified.

(a) Reflection from yellow box.

(b) Reflection from plastic funnel.

Fig. 7.6. Reflections (circled) during light stripe scanning in a simple domestic scene.

Figure 7.9 shows selected frames from the right camera during execution of the grasping task (see also *Multimedia Extensions*). Figure 7.9(a) shows the initial view, overlaid with tracking cues and the estimated pose of the box. Since the gripper is initially outside the field of view, EOL visual servoing commences using only kinematic measurements. Figure 7.9(b) shows the last stages of EOL control just before the end-effector fully enters the visual field. The large pose bias between the estimated pose (indicated by the wireframe overlay) and the actual pose of the gripper during EOL control in Figure 7.9(b)) is due to poor calibration of the camera to robot base transformation. This bias is overcome in Figure 7.9(c) after the gripper enters the field of view and visual measurements can be fused with kinematic measurements in the visual servo controller. At this stage, the end-effector has reached the intermediate set-point (noting that the grasp frame and target frame are aligned over the box) on approach to the planned grasp. Figure 7.9(d) shows the end-effector at the planned pose for grasping, and 7.9(e) shows the fingers in stable contact with the target object. Finally, the box is lifted in Figure 7.9(f) and the task is complete.

7.5 Experiment 2: Pouring Task

In this experiment, Metalman is required to grasp an interactively selected cup (filled with rice) and pour the contents into a bowl. The experimental arrangement for this task is shown in Figure 7.10. As before, no prior knowledge of the number and type of visible objects is assumed by the system. This experiment simulates the type of interaction that might be necessary when ambiguous or incomplete task specifications are encountered. For example, Metalman is given the command "Please pour the cup of rice into the bowl" and, after scanning the scene, finds that the supplied parameters do not sufficiently constrain the target. The robot consequently responds with: "Please indicate which cup I should use." In a practical application, the user may employ verbal or gestural interaction to provide additional parameters. The imple-

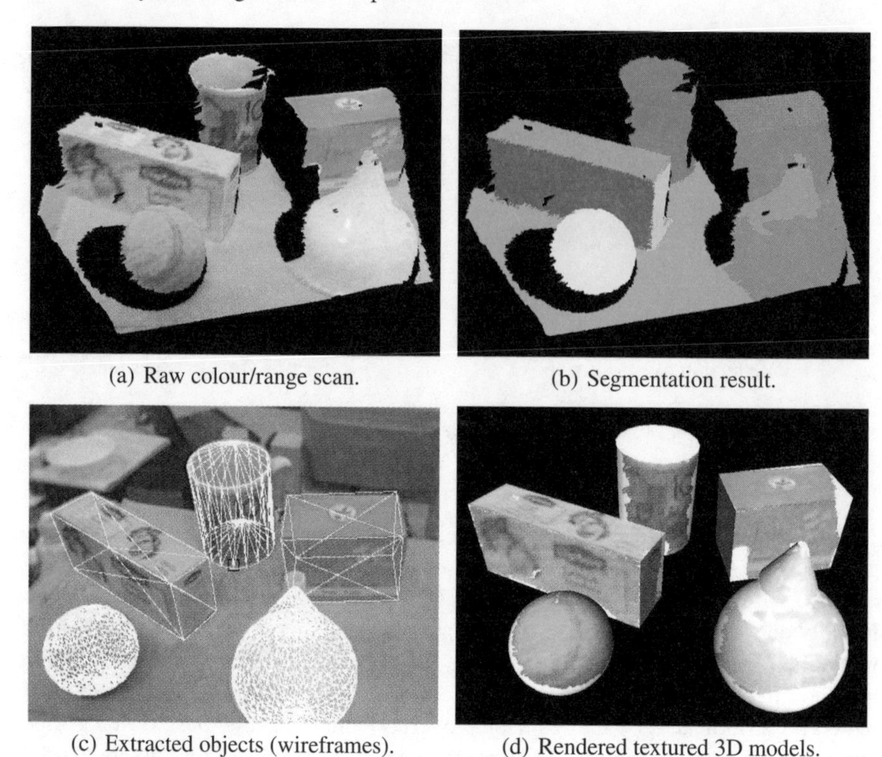

(a) Raw colour/range scan. (b) Segmentation result.

(c) Extracted objects (wireframes). (d) Rendered textured 3D models.

Fig. 7.7. Scene analysis for grasping task (see also *Multimedia Extensions*).

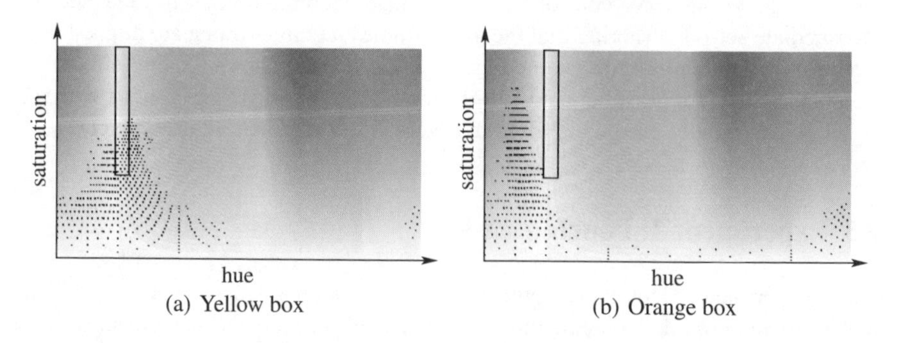

(a) Yellow box (b) Orange box

Fig. 7.8. Colour charts for identifying the yellow box. Texture pixels are plotted in black, and hue and saturation ranges (intensity not shown) for yellow are indicated by the rectangle.

mentation in this chapter employs a GUI, which may be applicable for tele-operated domestic robots (for example, see [146]).

(a) Tracking yellow box

(b) EOL servoing before gripper enters view

(c) Visual servoing to first set-point

(d) Planned grasp, before closing fingers

(e) Grasp closure

(f) Final pose at task completion

Fig. 7.9. Selected frames from grasping task (see *Multimedia Extensions* for a video of the complete experiment).

7.5.1 Implementation

The *sense-plan-action* cycle for this task resembles that of the previous experiment, and the details described in Section 7.4.1 also apply here. However, this experiment introduces interaction and more complex motion planning to pour the contents of the cup into the bowl, which are outlined below:

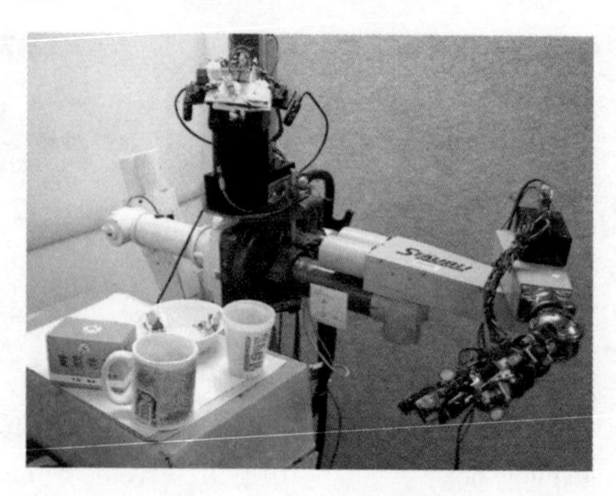

Fig. 7.10. Arrangement of objects for Experiment 2. Metalman is required to grasp one of the cups (selected interactively) and pour the contents into the bowl.

Interaction

Interaction is facilitated through a point-and-click GUI. Light stripe scanning and range data segmentation produce a list of cylindrical/conical objects, and the target bowl is detected as the object with the largest radius. The centroid of each remaining cup is projected onto the left image plane and rendered as a button over the possible target. The robot then waits for the user to select a cup by clicking on the appropriate button, and the selected object is tracked and grasped as described in Section 7.4.1.

Pouring Motion

The pouring motion is implemented as a set of predetermined set-points relative to the tracked position of the bowl and measured height of the cup. After lifting the cup, the first set-point centres the cup directly above the bowl, at an elevation which places the bottom of the cup slightly above the rim. The gripper is then rotated through 120 degrees and simultaneously lowered towards the bowl. The bowl is tracked at a rate of 2 Hz during the pouring motion to ensure accurate placement of the cup. However, the cup ceases to be tracked after the initial grasp (to reduce computational expense) and is simply assumed to maintain a stable pose relative to the gripper.

7.5.2 Results

Figure 7.11 shows the result of light stripe scanning and scene analysis for the pouring experiment. As in the previous experiment, secondary reflections caused by the smooth surfaces of the objects were readily rejected by the robust stereoscopic scanner, resulting in the colour/range scan in Figure 7.11(a). The segmented range map is shown in Figure 7.11(b). Convex objects extracted from the scene are overlaid as wireframe models in Figure 7.11(c) and rendered together in Figure 7.11(d). VRML models of the raw colour/range scan and extracted objects can be found in the *Multimedia extensions*. Again, these results highlight the accuracy and reliability of the

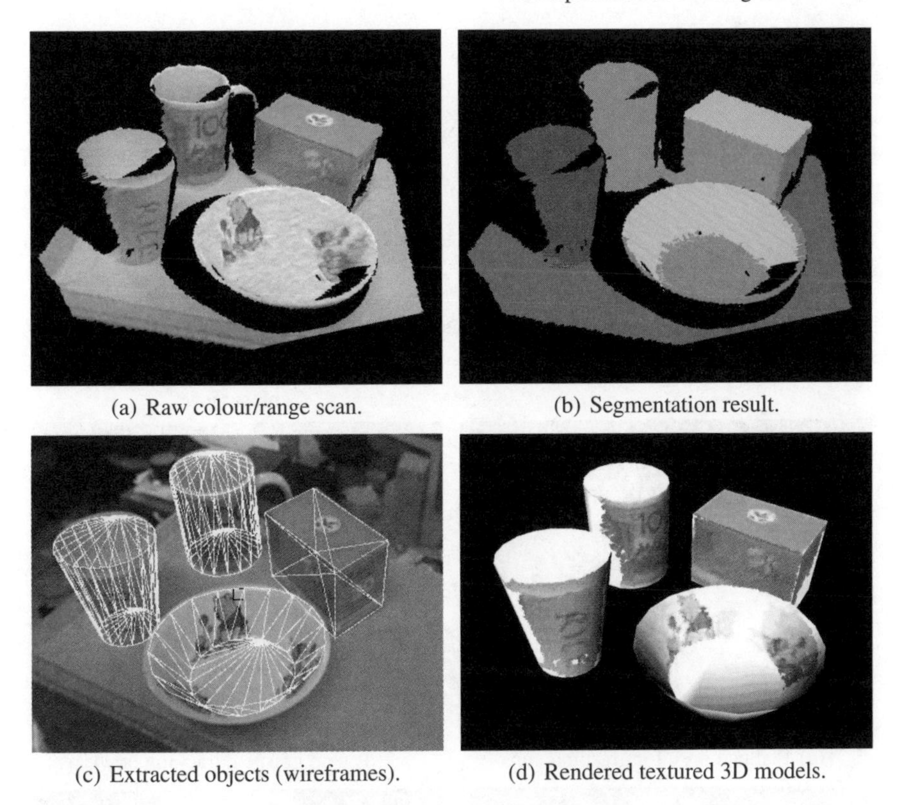

(a) Raw colour/range scan.

(b) Segmentation result.

(c) Extracted objects (wireframes).

(d) Rendered textured 3D models.

Fig. 7.11. Scene analysis for pouring task (see also *Multimedia Extensions*).

Fig. 7.12. Graphical interface for interaction: cup is selected by clicking on a rectangle.

proposed scanning/segmentation framework for modelling simple objects in unstructured scenes. However, segmentation required significantly greater processing time than the previous experiment (about one minute, in addition to the 30 seconds for

(a) Tracking cup, EOL servoing

(b) Hybrid visual servoing after gripper enters view

(c) Planned pose for grasping cup

(d) Cup grasped, tracking bowl

(e) Alignment of cup above bowl

(f) Successful completion of the task

Fig. 7.13. Selected frames from pouring task (see *Multimedia Extensions* for a video of the complete experiment).

range data acquisition) due to the large number of relatively expensive conical surfaces.

Scene analysis initially extracts three cylindrical/conical objects, and correctly identifies the bowl as having the largest radius. Figure 7.12 shows the GUI presented to the user to select from the remaining two objects. The target cup is selected by

Fig. 7.14. End-effector tracking loss due to background distraction (orange box).

clicking on the appropriate rectangle, in this case the cup marked "RICE" on the far left. It is also important to note from Figure 7.12 that the system has correctly discarded the box as a possible object of interest.

Selected frames from the right camera during execution of the task are shown in Figure 7.13 (see also *Multimedia Extensions*). Figure 7.13(a) shows the initial view; as in the previous experiment, the end-effector is initially unobservable, so visual servoing must commence using only kinematic feedback and the tracked pose of the cup. Figure 7.13(b) shows the captured image just after the end-effector fully enters the visual field and the controller initiates visual servoing using fused visual and kinematic measurements. The end-effector reaches the planned pose for grasping in Figure 7.13(c), and Figure 7.13(d) shows the cup grasped and lifted. At this point, tracking is terminated and the cup is assumed to maintain a stable pose relative to the end-effector. In any case, the significant deformation suffered by the cup during grasping would hinder model-based tracking. Tracking is initiated for the bowl, which is still only partially visible in Figure 7.13(d). Figure 7.13(e) shows the alignment of the cup above the bowl prior to pouring. Finally, the task is successfully completed in Figure 7.13(f), with the rice poured from the cup to the bowl.

After completing the task, Figure 7.14 shows that the system actually lost track of the gripper. This was the result of two compounding effects: only two LEDs were visible in the final pose, while the distracting orange box in the background gave the appearance of a third LED. However, it is important to note that bowl tracking was successful throughout the pouring manoeuvre, despite a poorly contrasting background and significant occlusions from the gripper and cup during servoing.

7.6 Experiment 3: Multi-sensor Synergy

Vision is usually considered one of the primary human senses, but it is easy to overlook the important role played by our other senses in accomplishing many simple tasks. For example, tactile and odour sensing also provide useful information for classifying objects that simply cannot be recovered from vision alone (consider the

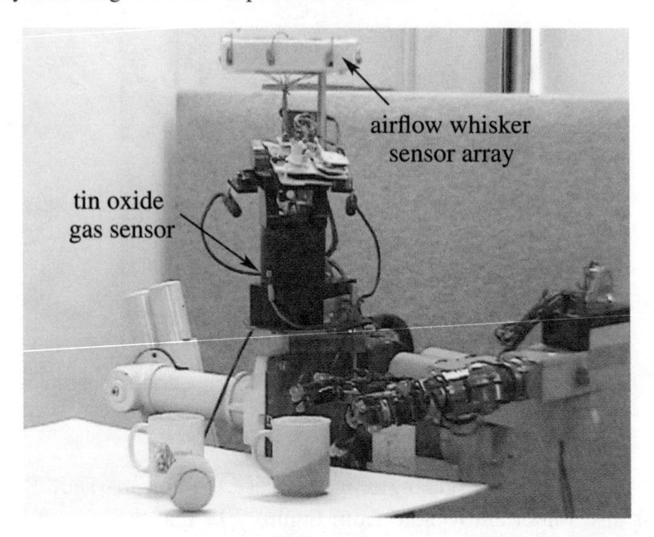

Fig. 7.15. Addition of airflow and odour sensors to Metalman for experiment 3.

difference between good and bad eggs!). Service robots of the future are likely to have a rich set of complementary sensing modalities going well beyond vision, which will enable them to maintain an increasingly complex world model and make better informed decisions.

This final experiment takes the first steps towards integrating vision with complementary sensors to complete a task that cannot be solved by vision alone: Metalman must locate and grasp a cup containing ethanol from among several objects on a table, including other cups. Ethanol is used in this case due to its high volatility, although one can imagine the analogous domestic task of distinguishing between a cup of coffee and a cup of tea. To detect the concentration of ethanol vapour, Metalman is equipped an electronic nose based on a tin oxide gas sensor. To help locate the source of the chemical plume, Metalman is also equipped with an airflow sensor capable of measuring both air speed and direction. Environmental airflow is generated by a domestic cooling fan driven from a variable transformer. Figure 7.15 shows the additional sensors mounted on Metalman. This collaborative work first appeared in [133].

7.6.1 Implementation

To accomplish this task, Metalman first analyses the scene using the light stripe sensor and object classification framework developed in Chapters 3 and 4 to locate all the cups on the table. The decision process is then triggered automatically when Metalman first detects the presence of ethanol, *ie.* when ethanol is poured onto one of the cups. The likelihood that each cup contains ethanol is evaluated over a period of time based on a recurrent stochastic process model, which takes air speed and direction, chemical concentration and the location of the cup into consideration. After

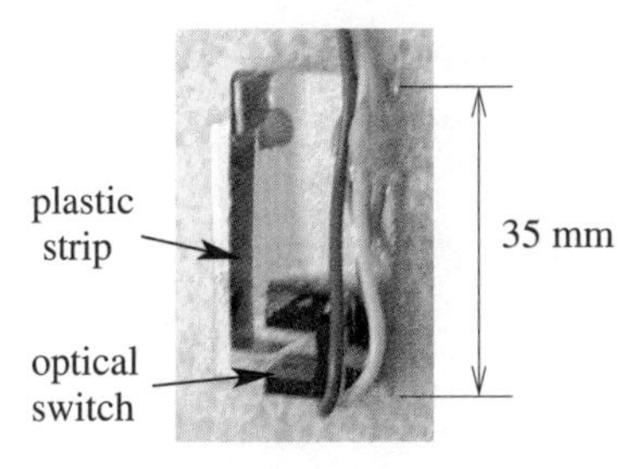

Fig. 7.16. Whisker sensor element for measuring airflow.

a fixed decision period, Metalman selects and grasps the cup with the highest likelihood of containing ethanol. Grasp planning and visual servoing are performed in the same manner as the previous experiments. However, object tracking was not used in this case as the workspace was limited and the camera experiences little motion. Once grasped, the presence of ethanol is confirmed by bringing the cup closer to the nose and detecting the associated increase in chemical concentration.

The following sections describe the operation of the airflow and chemical sensors, and the integration of sensor measurements in a recurrent stochastic model used to evaluate the likelihood of the presence of ethanol.

Airflow Sensor

The airflow sensor was developed by Russell and Purnamadjaja [132] and is based on the biologically inspired idea that airflow can be sensed as the disturbance of the fine hairs covering the skin. Based on this principle, Metalman uses an array of whisker sensors to measure both air speed and direction. Each sensor element is composed of a plastic film whisker and an optical switch, as shown in Figure 7.16. With careful alignment, the switch is capable of registering a 0.04 mm deflection in the tip of the whisker. Eight whiskers are arranged in a halo above the Biclops head (see Figure 7.15), and a microcontroller registers the number of vibrations from each sensor in a fixed period of time. The eight measurements form an airflow feature vector.

As mentioned above, a variable speed airflow is directed towards Metalman using a domestic cooling fan running from a variable transformer. The circulation of air around the whisker array depends critically on the location of both the fan and any obstacles deflecting the airflow. A simple learning procedure must be conducted to enable Metalman to predict the speed and direction of airflow in a given environment. The fan is moved to several known locations around the robot, and whisker measurements are recorded at each location for both high and low-speed airflow to form a set of training vectors. When Metalman is presented with an unknown airflow in the same environment, the whisker measurements are correlated with the training set and the highest result gives both the approximate direction and speed of airflow.

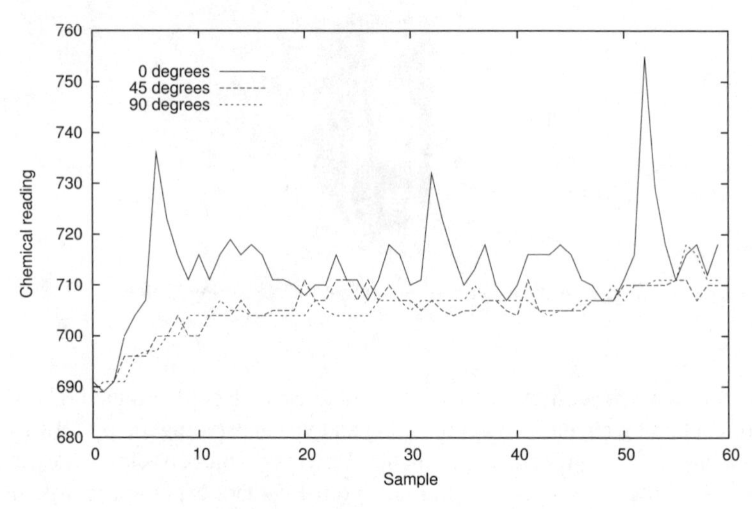

Fig. 7.17. Chemical sensor response to a mug of ethanol placed at different bearings to the direction of airflow.

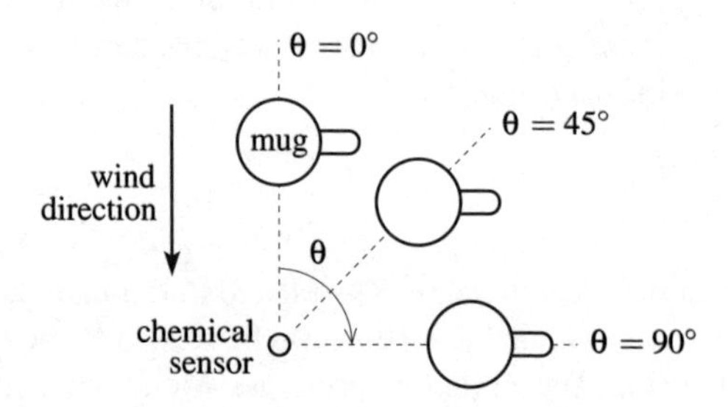

Fig. 7.18. Definition of bearing angle θ of mug with respect to wind direction.

Odour Sensor

Electronic noses are typically constructed as an array of chemical sensors, each sensitive to a broad but distinct range of chemicals (much like the different cones in the human retina that distinguish different colours). The set of responses to a particular chemical form a feature vector and classical pattern recognition techniques (commonly neural networks) can be applied to classify odours. Since classification is not required in this experiment, a single sensor is sufficient to detect the presence of ethanol. Metalman uses a Figaro tin oxide gas sensor, which presents a varying resistance as a non-linear function of chemical concentration.

Figure 7.17 shows the response of Metalman's odour sensor to a mug containing a teaspoon of ethanol. A slow airflow was aimed directly towards Metalman and the mug was placed at different bearings around the sensor. Figure 7.18 defines the bearing angle θ of the mug with respect to the direction of the wind. The initial transient in the response is due to the establishment of the stable down-wind chemical plume from the mug. The plot suggests that the location of the cup with respect to the direction of airflow is best characterized by the peak-to-peak variation in sensor measurements (after the initial transient), which in tun is determined by the location of the sensor within the down-wind plume.

Sensor Coordination

The measurements from vision, airflow and odour sensing arise from distinctly different physical processes and are orthogonal in the sense that measurements from one sensor cannot be recovered from any other. To integrate the sensors in a meaningful way, we construct a probabilistic framework in which measurements provide evidence supporting or disclaiming the hypothesis that a detected cup contains ethanol. A likelihood function is maintained for each cup to evaluate the support for this hypothesis. When asked to grasp the cup containing ethanol, Metalman naturally chooses the cup with the highest likelihood.

The likelihood function is based on the following observations. If no airflow is detected, the likelihood that a particular cup contains ethanol asymptotes towards the *a priori* likelihood (set arbitrarily to 0.5) since there is no evidence to suggest otherwise. Detecting a strong odour in the presence of airflow increases the likelihood that a cup upwind from the sensor (at zero bearing) contains ethanol. Conversely, detecting a weak odour decreases the likelihood that a cup upwind contains ethanol, since we expect a strong odour. The contents of a cup at 90 degrees bearing cannot be deduced (regardless of detected odour strength) since the chemical plume never reaches the sensor. Thus, the likelihood of a cup at 90 degrees always remains at the *a priori* likelihood. Cups at other bearings vary between these extreme cases.

To remove the effect of sensor noise and produce a time averaged result, the likelihood function is formulated as a first order, recurrent stochastic process. Let L^i_n represent the likelihood of the ith cup at the nth time step, θ^i represent the angle between the direction of the cup and the airflow with respect to the odour sensor, A^i represent the *a priori* likelihood that a cup contains ethanol and m model the time response of the likelihood function (for low-pass filtering). Then, based on the observations above, we can write the likelihood update equation as:

$$L^i_n = \begin{cases} L^i_{n-1} + (A^i - A^i \cos\theta^i - L^i_{n-1})/m & \text{if } \textit{low odour, airflow} \\ L^i_{n-1} + (A^i - L^i_{n-1})/m & \text{if } \textit{med. odour, airflow} \text{ or } \textit{no airflow} \\ L^i_{n-1} + (A^i + (1-A^i)\cos\theta^i - L^i_{n-1})/m & \text{if } \textit{high odour, airflow} \end{cases} \tag{7.1}$$

7.6.2 Results

Figure 7.19(a) shows the arrangement of objects, namely two cups and a tennis ball, and the initial configuration of Metalman for this experiment. The cooling fan is placed directly in front of Metalman so that the rightmost cup is almost directly upwind while the leftmost cup is away from the up-wind direction. Each trial began

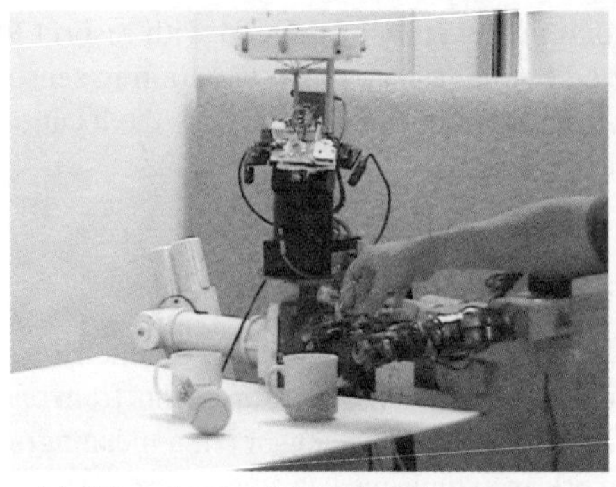

(a) Ethanol poured into the left mug, away from up-wind direction

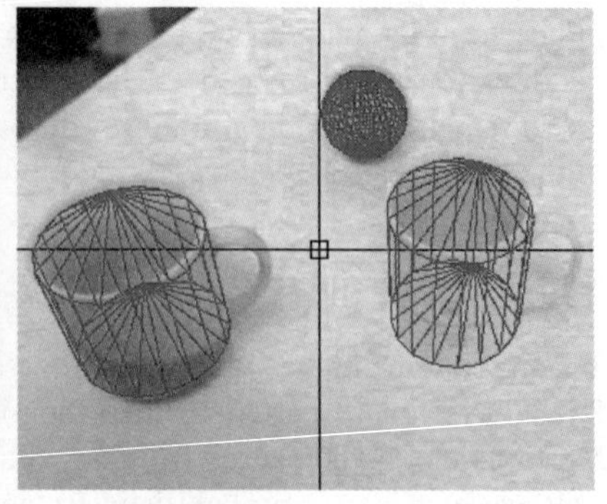

(b) Classified objects: two cups and a tennis ball (wireframe overlay)

Fig. 7.19. Experimental arrangement for experiment 3.

with a scan of the scene to recognize and locate the cups, as indicated by the wireframe models overlaid on Figure 7.19(b). The left and right cups are measured at bearings of 36 and 6 degrees respectively relative to the wind direction, while the ball is discarded as a possible source of ethanol. Ethanol was then poured into one of the cups to trigger the decision process prior to grasping. The experiment was conducted twice to test Metalman's ability to detect ethanol in each cup (see *Multimedia Extensions* for a video of the complete experiment).

For the first trial, ethanol was poured into the leftmost cup away from the airflow (see Figure 7.19(a)). Figure 7.20 shows the reading from the chemical sensor and the evolution of the likelihood functions for each cup. After the initial rise in chemical concentration above a fixed threshold at 55 seconds from the start of the trial, the system waits an additional minute while a stable plume is established. At 120 seconds, two likelihood functions are initialized and updated with each new odour and airflow reading. The small peak-to-peak variation in chemical concentration suggests that very little ethanol is present, and the likelihood functions for both cups drop below the *a priori* level of 0.5. The likelihood for the right cup drops furthest, since we would expect the strongest response if this upwind cup contained ethanol.

The likelihoods are evolved for a fixed period of 100 seconds, and the cup with the greatest likelihood, in this case the left cup, is selected for grasping. Figure 7.21 shows the left cup successfully grasped and brought closer to the odour sensor. Confirmation that the correct cup was selected is established by the large jump in the chemical sensor reading at about 270 seconds in Figure 7.20.

Figure 7.22(a) show the ethanol being poured into the right cup for the second trial in this experiment, and Figure 7.23 shows the chemical sensor reading and evolution of the likelihood functions. As before, the system waits for a minute after the initial rise in chemical concentration (at 40 seconds) before evolving the likelihood functions for each cup. In this case, the chemical readings show a high peak-to-peak variation indicating the presence of an upwind ethanol source. The likelihoods cor-

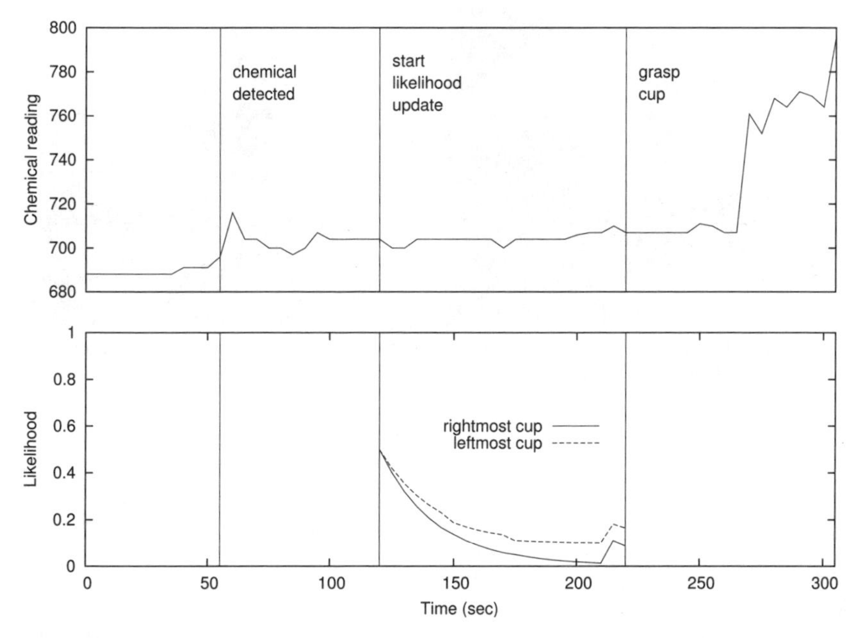

Fig. 7.20. Chemical sensor readings and evolution of likelihood function for both cups (ethanol in the left-hand cup).

(a) Visual servoing to planned grasp

(b) Correct cup grasped and lifted towards nose

Fig. 7.21. Successful completion of the detection and grasping task for left-hand cup.

respondingly rise above the *a priori* value of 0.5. After 100 seconds of evolution, the right cup is selected as the most likely to contain ethanol, being directly upwind. Figure 7.22 shows the cup successfully grasped and brought towards the nose. Again the

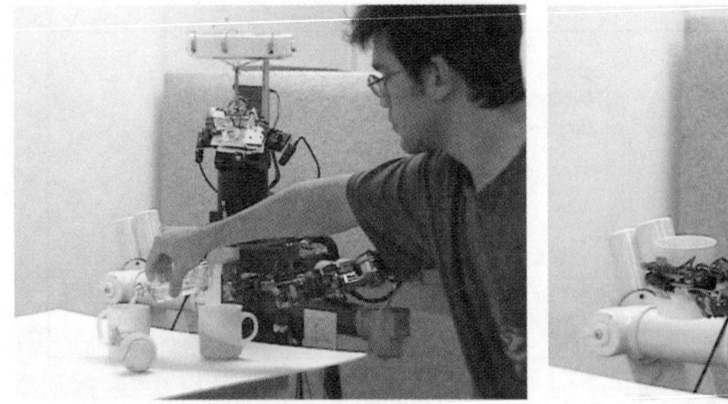

(a) Ethanol poured into the right mug, directly upwind

(b) Correct cup grasped and lifted towards nose

Fig. 7.22. Successful completion of the grasping task with ethanol in the rightmost cup.

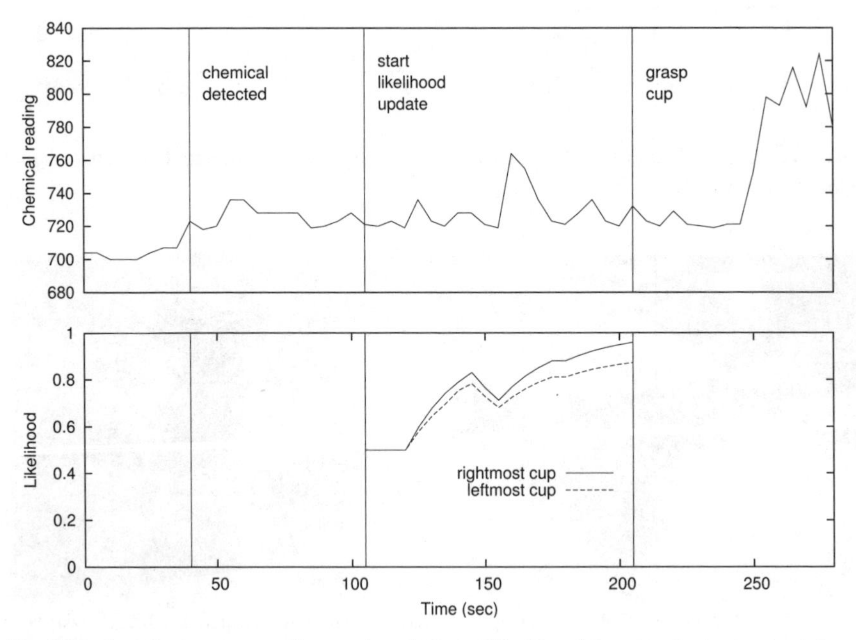

Fig. 7.23. Chemical sensor readings and evolution of likelihood function for both cups (ethanol in the rightmost cup).

presence of ethanol in the cup is confirmed by the jump in chemical concentration at about 250 seconds in Figure 7.23.

7.7 Discussion and Conclusions

In this chapter, three real-world tasks were experimentally implemented and tested to evaluate the performance of the proposed framework for visual perception and control for a hand-eye system. The tasks were based on supervisory commands and involved classifying and manipulating objects with which the robot had no prior experience. All of the methods developed in the preceding chapters, including stereo light stripe sensing, object modelling and classification, multi-cue tracking and hybrid position-based visual servoing were integrated in the implementation. Although the tasks were contrived to simplify scene analysis and avoid complex planning, their successful completion demonstrates the effectiveness of this framework for performing simple real-world tasks with a hand-eye system on a service robot.

The range scanning, segmentation and object modelling results were successful over many trials of each experiment. Importantly, significant secondary reflections of the laser stripe did not prevent the robust stereoscopic light stripe scanner from producing accurate, dense colour/range data. However, gripper tracking was occasionally distracted by background features despite the use of active cues and a global association algorithm, as demonstrated at the end of the second experiment. This problem was mainly compounded by the use of red LED markers, combined with the red light stripe scanner that requires all target objects in the scene to have some red component of colour. A possible solution to this problem may be the use of multiple coloured LEDs, which vary in colour across the gripper and between frames to greatly reduce the likelihood of a sustained distraction in the background. Alternatively, the robustness of multi-cue tracking in Chapter 5 suggests that a integration of different cues could improve gripper tracking. For example, the gripper could be augmented with distinctive line and texture markings in addition to multi-coloured LEDs.

In Chapter 1, real-time operation was noted as an important aspect of practical service robots. The execution time for the tasks presented in this chapter were approximately 80 s for the first experiment and 130 s for the second. The greatest proportion of time was consumed by light stripe scanning (25 s for bi-directional scans) and range data segmentation (40-60 s depending on the complexity of the scene). This is significantly longer than the time required for a human to perform a similar analysis. However, the current implementation is within an order of magnitude of what might be considered acceptable for a practical domestic robot (ie. a few seconds). With further optimization and high-speed hardware (such as CMOS cameras and digital signal processors), it is likely that the proposed methods could be realized in real-time using current technology.

Another aspect of real-time performance is the speed at which the arms are moved. The visual servo controller employed here has a deliberately low gain to avoid instabilities that arise from dynamic effects such as processing and actuation delays (see Appendix E). Compensation for these effects by developing more sophisticated visual servoing control laws is an active area of research (see, for example [27]), and will need to be solved before the robot can perform with the flexibility and speed for a human. In these experiments, the time required to move the gripper towards a planned grasp over a distance of 40 cm was approximately 10 s.

The above experiments avoided complex scene analysis and task planning by conveniently arranging simple objects. These concessions were also necessitated by the limitations of the experimental platform. In particular, the parallel-jaw configuration of the gripper eliminates the possibility of dextrous manipulation and reduces the range of objects that can be grasped. Furthermore, the kinematics of the Puma arm and lack of articulation about the waist of the robot significantly limits the available workspace. Since the kinematics were not explicitly considered in trajectory planning or control, careful arrangement of the objects was necessary to avoid joint limits. Clearly, more sophisticated planning (incorporating kinematics) must be adopted for a practical service robot.

Finally, the important problem of grasp verification was avoided for simplicity. Stability analysis could have been implemented using tracking to ensure the grasped object remains fixed in the hand. Similarly, tracking could be used to verify that objects reach their planned pose after each manoeuvre. If any component of the task fails, the system should be capable of planning steps to recover or alternatively reporting the mode of failure to the user. The above issues provide a starting point for future research to improve the performance and flexibility of hand-eye systems on practical service robots.

8

Summary and Future Work

8.1 Answering the Challenge of Service Robotics

Service robotics is currently a very active area of research, with significant progress being made in areas such as locomotion, interaction, learning and manipulation. The interest in service robots stems from the pressing needs of an aging society and diminishing workforce in many developed countries. Consequently, this book has focussed on developing manipulation skills that would be useful in applications such as aged care and construction work. Many of the important characteristics of practical service applications are largely neglected in current robotics research, including robust sensing in an unpredictable environment, tolerance to calibration errors, and minimal reliance on prior knowledge for tasks such as object recognition. Solving these problems contributes to the requirements of both high reliability and low cost that will drive the acceptance and success of service robots.

This book has developed a perception and control framework based on visual sensing to perform practical manipulation tasks with unknown objects in a domestic environment. Robustness to noise, interference and calibration errors at all stages from sensing through to actuation is a major theme of this framework. At the start of a new task, the robot acquires a dense 3D map of the workspace using the stereoscopic light stripe scanner described in Chapter 3. This sensor is capable of acquiring dense, accurate range data despite reflections, cross-talk or other sources of interference. Unknown objects are modelled as collections of data-driven geometric primitives (planes, cylinders, cones and spheres) fitted to the range data using the algorithms described in Chapter 4. This process relies heavily on a new algorithm to robustly classify the shape of local range patches in the presence of noise. The internal world model, consisting of textured polygonal models of extracted objects, is updated using the multi-cue 3D model-based tracking algorithm developed Chapter 5. Multi-cue integration allows objects to be tracked over a wide range of visual conditions, including clutter, partial occlusions, lighting variations and low contrasting backgrounds. Finally, the hybrid position-based visual servoing framework described in Chapter 6 was developed to control the end-effector with high tolerance to calibration errors, occlusions, and other distractions. Chapter 7 demonstrated the successful application of this above framework to several simple tasks involving unknown objects in a domestic setting.

Apart from forming the building blocks of a practical service robot, the techniques developed in this book have a wide range of applications outside robotics. As a prime example, light stripe scanners have been used in diverse fields including industrial inspection, CAD modelling, historical preservation and medical diagnosis. Robust light stripe sensing opens new possibilities such as capturing data "in the field", and capturing both colour and range while eliminating the need to prepare objects with matte white paint. Visual tracking is another component with a diverse range of applications including surveillance, sports and entertainment, human-machine interfaces and video compression. Multi-cue tracking has the potential to improve performance in all of these areas.

In Chapter 1, four main challenges associated with visual perception and control in a domestic environment were presented as the motivation for this research. It is therefore prudent to conclude this book by revisiting these challenges and summarizing how each has been addressed:

Imprecise task specifications and lack of prior knowledge

The experiments in Chapter 7 exemplified how service tasks are typically specified at a supervisory level, involving imprecisely described objects and manipulations. To further illustrate this point, these tasks were only specified by simple commands such as: "Please pass the yellow box" and "Please pour the cup of rice into the bowl." The robot was required to perform the tasks without necessarily having any prior experience with the objects involved, or any further clue regarding their location and identity. This lack of precise knowledge influenced all stages of sensing, planning and actuation in the framework developed in this book.

In an unpredictable environment with unknown objects, visual sensing cannot rely on the presence of suitable lighting conditions or particular visual cues. For low level sensing, active light stripe ranging was thus chosen over passive stereo, since the performance of the latter depends critically on the contents of the scene. Recognizing meaningful objects in the range data must be performed without precise models for objects present in the scene. This problem is addressed by modelling classes of objects as compound geometric primitives. Unknown objects are recognized by matching these models to data-driven geometric primitives segmented from the range data. Finally, object tracking cannot rely on the presence of intensity edges or unique textures and colours to distinguish objects, which causes conventional single-cue trackers to fail in the long term. Multi-cue tracking directly addresses this problem and thus caters for a wide range of targets and visual conditions.

The second experiment in Chapter 7 illustrated how imprecise task specifications might prevent a task from being successful executed. The cause in this case was the detection of two objects satisfying the task specifications. To successfully resolve this ambiguity, the robot asked for assistance using the current world model to compose a suitable query. This example illustrates the importance of human-machine interaction in robustly carrying out complex, cooperative domestic tasks.

Robust visual sensing in a cluttered environment

Data association, the task of identifying features in sensor data, is particularly problematic in a cluttered, dynamic and unpredictable domestic environment. This issue drove the development of the robust stereoscopic light stripe scanner described in Chapter 3. Conventional scanners rely on the brightness of the stripe exceeding all other features in the scene to solve the data association problem. This assumption is clearly violated under arbitrary lighting conditions and with the possibility of reflections and cross-talk. The stereoscopic light stripe scanner identifies the true stripe in the presence of interference by validating measurements using constraints provided by redundancy.

Object and gripper tracking suffer similar association problems, and two different solutions were demonstrated. Two critical techniques to improve object tracking are the use of multiple cues and the detection of new cues in each frame. The former process reduces the significance of single association errors, while the latter ensures that association errors do not persist for more than a single frame. Alternatively, gripper tracking is improved through the use of easily detected active cues and fusion with kinematic feedback. This latter approach fully exploits the additional constraints that can be imposed by the gripper being controlled by the robot. Fusion with kinematic measurements also overcomes the problem of losing visual feedback due to occlusions, which are a significant issue in an unpredictable, cluttered environment. The hybrid position-based controller implements pure EOL control or fusion of visual and kinematic measurements depending on the availability of visual information. Servoing is therefore possible over a wide range of conditions.

Robustness to operational wear and calibration errors

Accurate calibration of the sensors and mechanics in a service robot is hindered by the requirements of low weight for minimal power consumption, cheap sensors, low maintenance and operation in an unpredictable, hazardous environment. System parameters are likely to change with operating conditions, wear and accidents. Many of the techniques in this book were therefore developed to be self-calibrating rather than relying on accurate manual calibration, to meet the requirements of reliability and low maintenance.

Chapter 3 described an automatic procedure to calibrate the light stripe range sensor from the scan of an arbitrary non-planar target, which allows the scanner to be calibrated during normal operation. In Chapter 6 it was shown that mild verge and baseline errors in the stereo camera model can be approximated as a scalar transformation. This was exploited in visual servoing to compensate for camera calibration errors while tracking the pose of the gripper. The tracking filter also maintains an on-line estimate of the location of the robot base to compensate for kinematic calibration errors. Finally, the experimental results in Chapter 5 demonstrated that calibration errors cause camera motion to bias the estimated position of static objects, despite ego-motion compensation. Object tracking overcomes this bias to improve the success rate of grasps and other manipulations.

Real-time operation

To interact and work cooperatively with humans, service robots must perform tasks at a similar rate. Chapter 7 demonstrated that the algorithms presented in this book enable an experimental humanoid robot to perform tasks within the constraints of some practical applications, although performance is below what may be considered real-time. The current performance is achieved on a general purpose desktop PC without specialized signal processing hardware. In the case of light stripe scanning and visual servoing, real-time performance is limited by mechanical constraints rather than computing power. However, all other components are likely to benefit from more efficient optimizations, supplementary signal processing hardware and exploitation of the steady increase in general purpose computing power. The results of this book certainly suggest that a real-time implementation should be achievable with current technology.

8.2 A Foundation for the Future

The visual perception and control framework developed throughout this book is by no means complete. The development of practical service robots will require significant advancements in a variety of related fields including sensing, control, human-machine interaction, learning and artificial intelligence. We chose to examine the problem of visual perception and control for robotic grasping as it touches on many of these areas, or at least highlights the need for further research. To achieve meaningful results within the limited scope of this book, the real-world tasks in Chapter 7 were deliberately contrived to facilitate simple scene understanding, task planning (including grasp planning and obstacle avoidance) and user interaction, and remove the need for dextrous manipulation and cooperative manipulation with two (or more) arms. These missing skills are important components of a useful and flexible service robot, and remain the subject of intense research elsewhere.

The final experiment in Chapter 7 highlighted the effectiveness and utility of multi-sensor fusion in service applications. The integration of smell and airflow sensing with vision allowed the robot to perform a task that would be impossible with vision alone. One can imagine many other useful sensing modalities on a service robot platform, both human based such as tactile and vestibular sensing, and non-human based including sonar and GPS. Clearly, the reliability of grasping and manipulation would be improved with tactile and force feedback. Tactile sensing can provide information about otherwise hidden surfaces of an object, and supplement visual grasp planning with reactive grasp stabilization. Fragile objects, which are difficult to distinguish visually, could be handled more confidently with the aid of tactile feedback. Tactile and wrist force sensors would alert the robot to collisions, which could be used to both supplement obstacle detection and determine when an object has been placed appropriately.

Just as the sensing capabilities of service robots should not be limited to anthropomorphic modalities, neither should the configuration of the sensors be limited to the human blueprint. For example, one can imagine many task-specific cameras placed all around a robot for specialized functions such as user interaction, localization and visual servoing. The *eye-in-hand* camera configuration is already popular in the visual servoing, and an interesting future research direction for position-based visual servoing would be the fusion of both fixed camera and eye-in-hand configurations for hand-eye coordination.

A great deal of scope exists for the inclusion of learning algorithms to improve the performance and flexibility of the framework presented in this book. At a basic level, segmentation and object modelling rely heavily on several thresholds including patch size and surface fitting tolerances. While these were chosen heuristically for the experimental implementation in this book, parameter selection is particularly suited to the application of unsupervised learning. By roughly predicting the contents of the scene from the commanded task, learning could be conditioned by the consensus between perception and prediction. Parameter selection could be further refined with supplemental inputs such as an estimate of the variance of range noise for current visual conditions, which could also be learned autonomously. As the robot gains operational experience, the ability to interpret range data could steadily improve in response to changes in the environment.

At a higher level, object modelling and classification is another suitable area for the application of learning algorithms. The framework presented in chapter 4 classifies generic objects using predefined attributed graphs of geometric primitives. However, a domestic robot will almost certainly be required to recognize special classes of objects peculiar to its application. Through interaction and learning, new associations could be formed between collections of primitives and meaningful tokens (textual, vocal or otherwise). Attributed graphs could be created and refined over time through experience with multiple instances of each new class. This capability would greatly increase the flexibility of the service robot.

Ultimately, this research presented in this book forms only small component of a practical, autonomous service robot. In the immediate future, integrating the perception and control skills developed here with progressive research in verbal/gestural human-machine interaction would create a system that could already provide practical assistance in simple domestic tasks. With research continuing in a host of related areas including locomotion, localization and mapping, dextrous manipulation and learning, the future of service robots looks promising indeed.

A

Active Stereo Head Calibration

This appendix accompanies the discussion on active vision in Section 2.2.3, and describes a calibration procedure for the Biclops stereo head. Biclops is a commercially available robotic pan-tilt-verge camera platform that enables the viewpoint of the cameras to be controlled for active tracking. Biclops provides active vision for the experimental service robot detailed in Chapter 7 (see Figure 7.1).

When implementing model-based vision on an active vision head, accurate feedback of the pan, tilt and verge angles (see Figure 2.5) are important to determine the current camera projection matrices. The Biclops head reports these quantities in their measured units of encoder counts. Manual calibration is thus required to transform these values into meaningful units such as radians or degrees. Fortunately, a linear relationship exists between encoder count and angular position for the pan and tilt axes, and the scale factor can be calculated by recording the encoder value at the limits of motion for each axis and dividing by the factory specified angular range. Conversely, the verge mechanism of the Biclops head does not possess such as linear relationship between encoder counts and angular travel. As illustrated in Figure A.1, the Biclops verges the cameras from a single motor via a worm gear, nut and transmission arm with leaf spring hinges. As the nut travels along the worm gear, the cameras rotate with a symmetrical, non-linear change in verge angle and baseline. Manufacturer specifications for these relationships are not provided, but the verge angle can be calibrated manually using the image-based procedure described below.

The calibration procedure is based on placing a coloured target at varying known distances from the head and controlling the verge angle to drive the measured colour centroid to the centre of the left and right image planes. The smooth pursuit control laws given by equations (2.31)-(2.33), with ${}^L\mathbf{x} = ({}^Lx, {}^Ly)^\top$ and ${}^R\mathbf{x} = ({}^Rx, {}^Ry)^\top$ representing the measured colour centroid, are used to centre the target at each known distance d_i, and the verge encoder value at the completion of each trial is also recorded. Using a manually measured stereo baseline $2b$, the verge angle ν_i at the ith target position can be calculated as

$$\nu_i = \tan^{-1}(b/d_i) \tag{A.1}$$

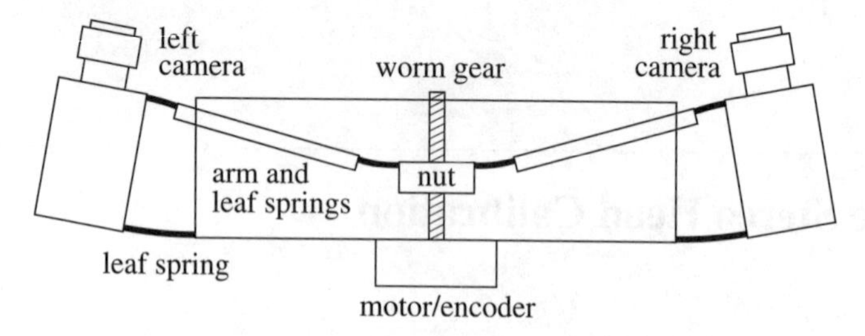

Fig. A.1. Verge mechanism for Biclops head (top view).

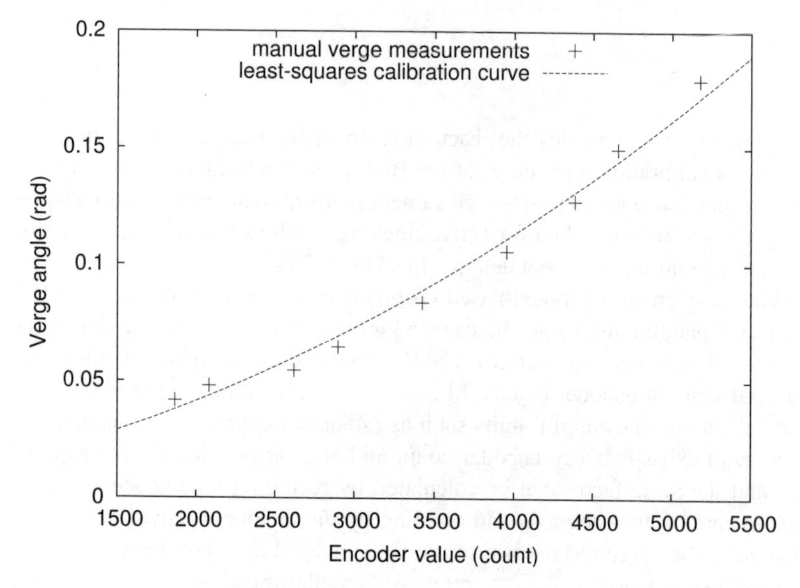

Fig. A.2. Calibration of verge angle.

Figure A.2 plots the measured verge angle (from equation (A.1)) against the corresponding encoder value, when this procedure was performed on the experimental Biclops head. Finally, an approximate quadratic calibration curve is calculated using a least squares fit, as shown by the dashed line in Figure A.2. While this approach only provides approximate verge and baseline calibration, Chapter 6 shows how the uncertainty in these parameters can be compensated during hybrid position-based visual servoing.

B

Light Stripe Validation and Reconstruction

This appendix presents detailed calculations of the theoretical results for the robust stereoscopic light stripe scanner described in Chapter 3.

B.1 Optimization of Light Plane Error Function

In Section 3.2.3, the following cost function is proposed to determine whether a pair of stereo measurements, ${}^L\mathbf{x}$ and ${}^R\mathbf{x}$, correspond to a point $\widehat{\mathbf{X}}$ on the light plane:

$$E = d^2({}^L\mathbf{x}, {}^L\mathrm{P}\widehat{\mathbf{X}}) + d^2({}^R\mathbf{x}, {}^R\mathrm{P}\widehat{\mathbf{X}}) \tag{B.1}$$

where $d(\mathbf{x}_1, \mathbf{x}_2)$ is the Euclidean distance between $\mathbf{x}_1$ and $\mathbf{x}_2$ and ${}^{L,R}\mathrm{P}$ are the projection matrices of the stereo cameras. The reconstruction $\widehat{\mathbf{X}}$ is constrained to the light plane $\boldsymbol{\Omega}$ by:

$$\boldsymbol{\Omega}^\top \widehat{\mathbf{X}} = 0 \tag{B.2}$$

The minimum error E^* and optimal reconstruction $\widehat{\mathbf{X}}$ for a given measurement pair is determined by the constrained optimization of equation (B.1) with respect to the constraint in (B.2). As noted, a direct optimization is analytically cumbersome, but the problem can be reduced to an unconstrained optimization by determining the direct relationship between projections ${}^L\hat{\mathbf{x}}$ and ${}^R\hat{\mathbf{x}}$ for points on the light plane.

The following formulation is based on the observation that $\widehat{\mathbf{X}}$ lies at the intersection of the light plane with the image ray back-projected through ${}^R\hat{\mathbf{x}}$ (or ${}^L\hat{\mathbf{x}}$). Plücker matrices provide a concise notation for the intersection of planes and lines (see [52]). If $\mathbf{A}$ and $\mathbf{B}$ represent the homogeneous vectors of two points on a line, the Plücker matrix L describing the line is

$$\mathrm{L} = \mathbf{A}\mathbf{B}^\top - \mathbf{B}\mathbf{A}^\top \tag{B.3}$$

Then, the intersection $\mathbf{X}$ of a plane $\boldsymbol{\Omega}$ and the line described by L is simply

$$\mathbf{X} = \mathrm{L}\boldsymbol{\Omega} = (\mathbf{A}\mathbf{B}^\top - \mathbf{B}\mathbf{A}^\top)\boldsymbol{\Omega} \tag{B.4}$$

Now, the Plücker matrix $\mathbf{L}_R$ for the back-projection of ${}^R\hat{\mathbf{x}}$ can be constructed from two known points on the ray: $\mathbf{C}_R$ and ${}^R\mathrm{P}^{+R}\hat{\mathbf{x}}$, where ${}^R\mathrm{P}^{+}$ is the pseudo-inverse of the camera projection matrix ${}^R\mathrm{P}$, given by equation (2.35). Applying these to equation (B.3), the Plücker matrix L_R for the back-projection of ${}^R\hat{\mathbf{x}}$ is

$$\mathrm{L}_R = \mathbf{C}_R({}^R\mathrm{P}^{+R}\hat{\mathbf{x}})^\top - ({}^R\mathrm{P}^{+R}\hat{\mathbf{x}})\mathbf{C}_R^\top \tag{B.5}$$

The intersection of L_R with the laser plane $\boldsymbol{\Omega}$, can now be expressed using equation (B.4) as:

$$\widehat{\mathbf{X}} = \mathrm{L}_R\boldsymbol{\Omega} = [\mathbf{C}_R({}^R\mathrm{P}^{+R}\hat{\mathbf{x}})^\top - ({}^R\mathrm{P}^{+R}\hat{\mathbf{x}})\mathbf{C}_R^\top]\boldsymbol{\Omega} \tag{B.6}$$

Equation (B.6) is the result quoted in equation (3.9) of Section 3.2.3. Finally, the left projection ${}^L\hat{\mathbf{x}}$ corresponding to ${}^R\hat{\mathbf{x}}$ is obtained by projecting $\widehat{\mathbf{X}}$, given by equation (B.6), via ${}^L\mathrm{P}$:

$$\begin{aligned}{}^L\hat{\mathbf{x}} &= {}^L\mathrm{P}\widehat{\mathbf{X}}\\ &= {}^L\mathrm{P}[\mathbf{C}_R({}^R\mathrm{P}^{+R}\hat{\mathbf{x}})^\top - ({}^R\mathrm{P}^{+R}\hat{\mathbf{x}})\mathbf{C}_R^\top]\boldsymbol{\Omega}\\ &= {}^L\mathrm{P}\mathbf{C}_R({}^R\mathrm{P}^{+R}\hat{\mathbf{x}})^\top\boldsymbol{\Omega} - {}^L\mathrm{P}({}^R\mathrm{P}^{+R}\hat{\mathbf{x}})\mathbf{C}_R^\top\boldsymbol{\Omega}\end{aligned} \tag{B.7}$$

Using the identity $({}^R\mathrm{P}^{+R}\hat{\mathbf{x}})^\top\boldsymbol{\Omega} = \boldsymbol{\Omega}^\top({}^R\mathrm{P}^{+R}\hat{\mathbf{x}})$ (since both sides are scalar), and noting that $(\mathbf{C}_R^\top\boldsymbol{\Omega})$ is scalar, the common factors are collected to simplify the above expression to

$$\begin{aligned}{}^L\mathbf{x} &= {}^L\mathrm{P}(\mathbf{C}_R\boldsymbol{\Omega}^\top)({}^R\mathrm{P}^{+R}\hat{\mathbf{x}}) - {}^L\mathrm{P}(\mathbf{C}_R^\top\boldsymbol{\Omega})({}^R\mathrm{P}^{+R}\hat{\mathbf{x}})\\ &= \left({}^L\mathrm{P}[\mathbf{C}_R\boldsymbol{\Omega}^\top - (\mathbf{C}_R^\top\boldsymbol{\Omega})\mathrm{I}]{}^R\mathrm{P}^{+}\right)\hat{\mathbf{x}}_R\end{aligned} \tag{B.8}$$

Equation (B.8) is the desired relationship between projections ${}^L\hat{\mathbf{x}}$ and ${}^R\hat{\mathbf{x}}$ of points on the light plane $\boldsymbol{\Omega}$, and is the result quoted in equation (3.10) of Section 3.2.3. Finally, the error function in equation (B.1) can be written as

$$E = d^2({}^L\mathbf{x}, \mathrm{H}{}^R\hat{\mathbf{x}}) + d^2({}^R\mathbf{x}, {}^R\hat{\mathbf{x}}) \tag{B.9}$$

with $\mathrm{H} = {}^L\mathrm{P}[\mathbf{C}_R\boldsymbol{\Omega}^\top - (\mathbf{C}_R^\top\boldsymbol{\Omega})\mathrm{I}]{}^R\mathrm{P}^{+}$ and the minimum error can be found by an unconstrained optimization of equation (B.9) with respect to the projection ${}^R\hat{\mathbf{x}}$.

B.2 Optimal Reconstruction for Rectilinear Stereo and Pin-Hole Cameras

Equation (B.8) will now be evaluated for the case of pin-hole cameras in a rectilinear stereo configuration. The camera centres $\mathbf{C}_{L,R}$ and projection matrices ${}^{L,R}\mathrm{P}$ for rectilinear pin-hole cameras are given by (see Section 2.2.3)

$$\mathbf{C}_{L,R} = (\mp b,\ 0, 0,\ 1)^\top \tag{B.10}$$

$${}^{L,R}\mathrm{P} = \begin{pmatrix} f & 0 & 0 & \pm fb \\ 0 & f & 0 & 0 \\ 0 & 0 & 1 & 0 \end{pmatrix} \tag{B.11}$$

where the top sign is taken for L and the bottom sign for R. Substituting ${}^R\mathrm{P}$ into equation (2.35), the pseudo-inverse for the right projection matrix is

$$\begin{aligned} {}^R\mathrm{P}^+ &= {}^R\mathrm{P}^\top({}^R\mathrm{P}{}^R\mathrm{P}^\top)^{-1} \\ &= \begin{pmatrix} 1/[f(1+b^2)] & 0 & 0 \\ 0 & 1/f & 0 \\ 0 & 0 & 1 \\ -b/[f(1+b^2)] & 0 & 0 \end{pmatrix} \end{aligned} \tag{B.12}$$

Next, the term $[\mathbf{C}_R\mathbf{\Omega}^\top - (\mathbf{C}_R^\top\mathbf{\Omega})\mathrm{I}]$ in equation (B.8) evaluates to:

$$\mathbf{C}_R\mathbf{\Omega}^\top - (\mathbf{C}_R^\top\mathbf{\Omega})\mathrm{I} = \begin{pmatrix} -D & bB & bC & bD \\ 0 & -(Ab+D) & 0 & 0 \\ 0 & 0 & -(Ab+D) & 0 \\ A & B & C & -Ab \end{pmatrix} \tag{B.13}$$

Finally, multiplying the matrices in equations (B.11), (B.13) and (B.12), equation (B.8) is evaluated as:

$${}^L\hat{\mathbf{x}} = \begin{pmatrix} Ab-D & 2Bb & 2Cbf \\ 0 & -(Ab+D) & 0 \\ 0 & 0 & -(Ab+D) \end{pmatrix} {}^R\hat{\mathbf{x}} \tag{B.14}$$

which is the result quoted in equation (3.12) in Section 3.2.4. Evaluating equation (B.14) with ${}^L\hat{\mathbf{x}} = ({}^L\hat{x}, {}^L\hat{y}, {}^L\hat{w})^\top$ and ${}^R\hat{\mathbf{x}} = ({}^R\hat{x}, {}^R\hat{y}, 1)^\top$ gives

$$\begin{aligned} \begin{pmatrix} {}^L\hat{x} \\ {}^L\hat{y} \\ {}^L\hat{w} \end{pmatrix} &= \begin{pmatrix} Ab-D & 2Bb & 2Cbf \\ 0 & -(Ab+D) & 0 \\ 0 & 0 & -(Ab+D) \end{pmatrix} \begin{pmatrix} {}^R\hat{x} \\ {}^R\hat{y} \\ 1 \end{pmatrix} \\ &= \begin{pmatrix} (Ab-D){}^R\hat{x} + 2Bb{}^R\hat{y} + 2Cbf \\ -(Ab+D){}^R\hat{y} \\ -(Ab+D) \end{pmatrix} \end{aligned} \tag{B.15}$$

Expressed in inhomogeneous coordinates, the relationship between ${}^L\hat{\mathbf{x}}$ and ${}^R\hat{\mathbf{x}}$ is

$${}^L\hat{x} = -\frac{(Ab-D){}^R\hat{x} + 2Bb{}^R\hat{y} + 2Cbf}{Ab+D} \tag{B.16}$$

$${}^L\hat{y} = {}^R\hat{y} \tag{B.17}$$

which is the result quoted in equations (3.13)-(3.14) in Section 3.2.4.

Now, with ${}^{L,R}\hat{y} = y$ where $y = {}^Ly = {}^Ry$ is the height of the scan-line on which ${}^L\mathbf{x}$ and ${}^R\mathbf{x}$ are measured (see the discussion in Section 3.2.4), the error function in equation (B.9) can be evaluated from equations (B.16)-(B.17) as:

$$\begin{aligned} E &= d^2({}^L\mathbf{x}, {}^L\hat{\mathbf{x}}) + d^2({}^R\mathbf{x}, {}^R\hat{\mathbf{x}}) \\ &= ({}^Lx - {}^L\hat{x})^2 + ({}^Ly - {}^L\hat{y})^2 + ({}^Rx - {}^R\hat{x})^2 + ({}^Ry - {}^R\hat{y})^2 \\ &= \left({}^Lx + \frac{Ab-D}{Ab+D}{}^R\hat{x} + \frac{2Bb}{Ab+D}y + \frac{2Cb}{Ab+D}f\right)^2 + ({}^Rx - {}^R\hat{x})^2 \\ &= ({}^Lx + \alpha{}^R\hat{x} + \beta y + \gamma f)^2 + ({}^Rx - {}^R\hat{x})^2 \end{aligned} \tag{B.18}$$

where the last line makes the change of variables

$$\alpha = (Ab - D)/(Ab + D) \tag{B.19}$$
$$\beta = 2Bb/(Ab + D) \tag{B.20}$$
$$\gamma = 2Cb/(Ab + D) \tag{B.21}$$

Equations (B.18)-(B.21) are the result quoted in equations (3.15)-(3.18) in Section 3.2.4.

Equation (B.18) expresses the image plane error for points on the light plane as a function of a *single* variable, ${}^{R}\hat{x}$. Optimization now proceeds using standard techniques, setting $\frac{dE}{d^{R}\hat{x}}$ to zero:

$$\begin{aligned}\frac{dE}{d^{R}\hat{x}} &= 2\alpha({}^{L}x + \alpha^{R}\hat{x} + \beta y + \gamma f) - 2({}^{R}x - {}^{R}\hat{x}) \\ &= 2^{R}\hat{x}(\alpha^2 + 1) + 2[\alpha({}^{L}x + \beta y + \gamma f) - {}^{R}x] \\ &= 0\end{aligned}$$

Solving for ${}^{R}\hat{x}$ gives the optimal projection ${}^{R}\hat{x}^{*}$:

$${}^{R}\hat{x}^{*} = \frac{{}^{R}x - \alpha({}^{L}x + \beta y + \gamma f)}{\alpha^2 + 1} \tag{B.22}$$

Substituting equation (B.22) into (B.18) gives the minimum error E^{*} for the optimal reconstruction:

$$\begin{aligned}E^{*} &= \left(\frac{\alpha^{R}x + ({}^{L}x + \beta y + \gamma f)}{\alpha^2 + 1}\right)^2 + \left(\frac{\alpha^{2R}\hat{x} + \alpha({}^{L}x + \beta y + \gamma f)}{\alpha^2 + 1}\right)^2 \\ &= \frac{1}{(\alpha^2+1)^2}(\alpha^{R}x + {}^{L}x + \beta y + \gamma f)^2 + \frac{\alpha^2}{(\alpha^2+1)^2}(\alpha^{R}\hat{x} + {}^{L}x + \beta y + \gamma f)^2 \\ &= \frac{({}^{L}x + \alpha^{R}x + \beta y + \gamma f)^2}{\alpha^2 + 1}\end{aligned} \tag{B.23}$$

Equations (B.22) and (B.23) are the results quoted in equations (3.25) and (3.26) in Section 3.2.4. Finally, the optimal 3D reconstruction $\widehat{\mathbf{X}}^{*}$ is recovered by substituting ${}^{R}\hat{\mathbf{x}}^{*}$ into equation (B.6). The Plücker matrix L_R for the back-projected ray through ${}^{R}\hat{\mathbf{x}}^{*} = ({}^{R}\hat{x}^{*}, y)^{\top}$ is evaluated by substituting equations (B.10)-(B.12) into (B.5):

$$\begin{aligned}\mathrm{L}_R &= \mathbf{C}_R({}^{R}\mathrm{P}^{+R}\hat{\mathbf{x}}^{*})^{\top} - ({}^{R}\mathrm{P}^{+R}\hat{\mathbf{x}}^{*})\mathbf{C}_R^{\top} \\ &= \begin{pmatrix} 0 & by/f & b & -{}^{R}\hat{x}^{*}/f \\ -by/f & 0 & 0 & -y/f \\ -b & 0 & 0 & -1 \\ {}^{R}\hat{x}^{*}/f & y/f & 1 & 0 \end{pmatrix}\end{aligned} \tag{B.24}$$

Multiplying L_R by the plane parameters $\mathbf{\Omega} = (A, B, C, D)^{\top}$ and substituting the result in equation (B.22) gives the optimal reconstruction $\widehat{\mathbf{X}}^{*}$ in homogeneous coordinates:

$$\begin{aligned}\widehat{\mathbf{X}}^* &= \mathrm{L}_R \boldsymbol{\Omega} \\ &= \begin{pmatrix} Bby/f + Cb - D^R\hat{x}^*/f \\ -(Ab+D)y/f \\ -(Ab+D) \\ A^R\hat{x}^*/f + By/f + C \end{pmatrix}\end{aligned} \tag{B.25}$$

The x-coordinate of $\widehat{\mathbf{X}}^*$ in inhomogeneous notation is calculated by dividing the first row of equation (B.25) by the fourth row:

$$\widehat{X}^* = \frac{Bby + Cbf - D^R\hat{x}^*}{A^R\hat{x}^* + By + Cf} \tag{B.26}$$

Multiplying the top and bottom row by $2b/(Ab+D)$, and then making the change of variables in equations (B.19)-(B.21), noting that $(\alpha+1) = 2Ab/(Ab+D)$ and $(\alpha-1) = -2D/(Ab+D)$, gives $\widehat{X}^*$ in terms of α, β and γ:

$$\widehat{X}^* = \frac{b[\beta y + \gamma f + (\alpha-1)^R\hat{x}^*]}{(\alpha+1)^R\hat{x}^* + \beta y + \gamma f} \tag{B.27}$$

The expressions for $\widehat{Y}^*$ and $\widehat{Z}^*$ follow similarly

$$\widehat{Y}^* = \frac{-2by}{(\alpha+1)^R\hat{x}^* + \beta y + \gamma f} \tag{B.28}$$

$$\widehat{Z}^* = \frac{-2bf}{(\alpha+1)^R\hat{x}^* + \beta y + \gamma f} \tag{B.29}$$

Equation (B.29) is used to derive the limits of valid image plane measurements in Section 3.2.6. Finally, substituting ${}^R\hat{x}^*$ from equation (B.22) into equations (B.27)-(B.29), the inhomogeneous coordinates of the optimal reconstruction $\widehat{\mathbf{X}}^*$ can be written as a function of the image plane measurements ${}^L\mathbf{x} = ({}^Lx, y)^\top$ and ${}^R\mathbf{x} = ({}^Rx, y)^\top$, and system parameters α, β, γ, b and f:

$$\widehat{X}^* = \frac{[(\alpha-1)(\alpha^Lx - {}^Rx) - (\alpha+1)(\beta y + \gamma f)]b}{(\alpha+1)(\alpha^Lx - {}^Rx) + (\alpha-1)(\beta y + \gamma f)} \tag{B.30}$$

$$\widehat{Y}^* = \frac{2by(\alpha^2+1)}{(\alpha+1)(\alpha^Lx - {}^Rx) + (\alpha-1)(\beta y + \gamma f)} \tag{B.31}$$

$$\widehat{Z}^* = \frac{2bf(\alpha^2+1)}{(\alpha+1)(\alpha^Lx - {}^Rx) + (\alpha-1)(\beta y + \gamma f)} \tag{B.32}$$

which is the result quoted in equations (3.28)-(3.30) in Section 3.2.4.

C

Iterated Extended Kalman Filter

This Appendix presents an overview of the *Iterated Extended Kalman Filter* (IEKF) equations, which are used in Chapters 5 and 6 for model-based object tracking. A detailed treatment of Kalman filter theory and the IEKF can be found in [8] and [74]. The purpose of the Kalman filter is to estimate the state and error covariance matrix of a linear dynamic system from measurements with additive noise, and the IEKF provides a near-optimal solution when the measurement model is non-linear. The Kalman filter assumes that the measurement and system noise are Gaussian distributed, zero mean and white with known covariance. When these assumptions are not satisfied, such as for biased measurements, the filter equations under-estimate the true state covariance.

Let $\mathbf{x}(k)$ represent the state vector (the variables we wish to estimate), $\mathrm{P}(k)$ represent the covariance matrix describing the uncertainty in the state, and $\mathbf{y}(k)$ represent the measurements on a linear dynamic system at sample time k. Assuming the system has no inputs, the evolution of the state can be described by a discrete time *state transition equation* (also known as the *dynamic model*):

$$\mathbf{x}(k+1) = \mathrm{F}\mathbf{x}(k) + \mathbf{v}(k) \tag{C.1}$$

where F is the state transition matrix and $\mathbf{v}(k)$ is the process noise, which absorbs unmodelled dynamics. The measurements $\mathbf{y}(k)$ are related to the state by a *measurement model* of the form:

$$\mathbf{y}(k+1) = \mathrm{H}(k+1)\mathbf{x}(k+1) + \mathbf{w}(k+1) \tag{C.2}$$

where H represents the measurement function, and $\mathbf{w}(k)$ is the measurements noise. For non-linear sensors, such as the pin-hole camera used throughout this book, the measurement model takes the form

$$\mathbf{y}(k+1) = \mathbf{h}(k+1, \mathbf{x}(k+1)) + \mathbf{w}(k+1) \tag{C.3}$$

where $\mathbf{h}(k, \mathbf{x})$ is the vector-valued non-linear measurement function describing the sensor model. As already noted, the IEKF assumes $\mathbf{v}(k)$ and $\mathbf{w}(k)$ are white, zero mean, Gaussian distributed noise sources. The covariance matrices for the process noise $Q(k)$ and measurement noise $R(k)$ are assumed to be known, and are given by

$$Q(k) = E[\mathbf{v}(k)\mathbf{v}^\top(k)] \tag{C.4}$$

$$R(k) = E[\mathbf{w}(k)\mathbf{w}^\top(k)] \tag{C.5}$$

where $E[\cdot]$ denotes expectation. When the state and measurement noise components are assumed to be independent, as is the case for the systems in this book, R(k) and Q(k) are diagonal.

Let $\hat{\mathbf{x}}(k|k)$ represent the estimated state at time k, and $\mathbf{y}(k+1)$ represent a new set of measurements observed at time $k+1$. The IEKF algorithm estimates the new state $\hat{\mathbf{x}}(k+1|k+1)$ as a weighted mean of the state predicted by the system dynamics and the new measurements, with the filter weight calculated from the state error and measurement error covariances. In the first step of the algorithm, a predicted state vector $\hat{\mathbf{x}}(k+1|k)$ and covariance matrix $P(k+1|k)$ are calculated as

$$\hat{\mathbf{x}}(k+1|k) = F\hat{\mathbf{x}}(k|k) \tag{C.6}$$

$$P(k+1|k) = FP(k|k)F^\top + Q(k) \tag{C.7}$$

The predicted state is then updated using the new measurements, which is an iterated process in the IEKF to solve the non-linear measurement equations. Let $\mathbf{n}_i$ represent the updated state estimate at the ith iteration, with $\mathbf{n}_0 = \hat{\mathbf{x}}(k+1|k)$. To calculate the filter weight, the measurement function $\mathbf{h}(k+1)$ is replaced by a linear approximation $M_i(k+1)$, operating about the current updated state:

$$M_i(k+1, \mathbf{n}_i) = \left.\frac{\partial \mathbf{h}(k+1, \mathbf{x})}{\partial \mathbf{x}}\right|_{\mathbf{x}=\mathbf{n}_i} \tag{C.8}$$

Then, the filter weight $K_i(k+1)$ at the ith iteration is given by

$$K_i(k+1) = P(k+1|k)M_i^\top(k+1)[M_i(k+1)P(k+1|k)M_i^\top(k+1) + R(k+1)]^{-1} \tag{C.9}$$

Using the non-linear sensor model, the measurements associated with the current state estimate are predicted as $\mathbf{h}(k+1, \mathbf{n}_i)$ and compared to the actual measurements $\mathbf{y}(k+1)$ to form an observation error $\mathbf{e}(k+1)$:

$$\mathbf{e}(k+1) = \mathbf{y}(k+1) - \mathbf{h}(k+1, \mathbf{n}_i) \tag{C.10}$$

Finally, a new estimate $\mathbf{n}_{i+1}$ of the state is recovered as the weighted mean of the predicted state $\hat{\mathbf{x}}(k+1|k)$ and the observation error $\mathbf{e}(k+1)$ ([74], page 279):

$$\mathbf{n}_{i+1} = \hat{\mathbf{x}}(k+1|k) + K_i(k+1)\{\mathbf{e}(k+1) - M_i(k+1)[\hat{\mathbf{x}}(k+1|k) - \mathbf{n}_i]\} \tag{C.11}$$

Equations (C.8)-(C.11) are iterated until successive state estimates $\mathbf{n}_i$ converge according to suitable criteria. In the current implementation, the system is deemed to have converged when the maximum absolute error between successive estimates of all state variables is below a threshold c_{th}:

$$\max_j |n_{j,i+1} - n_{j,i}| < c_{th} \tag{C.12}$$

where $n_{j,i}$ is the jth element of vector $\mathbf{n}_i$. The final result is taken as the updated state estimate $\hat{\mathbf{x}}(k+1|k+1)$. Finally, the state covariance is updated from the predicted covariance (equation (C.7)) by

$$\mathrm{P}(k+1|k+1) = [\mathrm{I} - \mathrm{K}(k+1)\mathrm{M}(k+1)]\mathrm{P}(k+1|k) \tag{C.13}$$

In practice, the Joseph form covariance update is used:

$$\mathrm{P}(k+1|k+1) = \mathrm{A}(k+1)\mathrm{P}(k+1|k)\mathrm{A}^\top(k+1) + \mathrm{K}(k+1)\mathrm{R}(k+1)\mathrm{K}^\top(k+1) \tag{C.14}$$

where

$$\mathrm{A}(k+1) = \mathrm{I} - \mathrm{K}(k+1)\mathrm{M}(k+1) \tag{C.15}$$

which is less sensitive to numerical round-off errors ([8], pg. 216).

D

Stereo Reconstruction Error Models

This appendix presents error models describing the uncertainty in reconstructing a single point from stereo measurement, and similarly estimating the pose of an object modelled by multiple points. The error models are used in Chapter 6 for hybrid position-based visual servoing.

D.1 Optimal Reconstruction of a Single Point

This section derives the optimal stereo reconstruction $\widehat{\mathbf{X}}$ of a point from measurement ${}^{L}\mathbf{x}$ and ${}^{R}\mathbf{x}$ on the left and right image planes. Assuming pin-hole cameras in a rectilinear configuration, the projections of $\widehat{\mathbf{X}}$ on the stereo image planes are given by

$$ {}^{L,R}\hat{\mathbf{x}} = {}^{L,R}\mathrm{P}\widehat{\mathbf{X}} \tag{D.1} $$

where ${}^{L,R}\mathrm{P}$ are the projection matrices of the left and right cameras, given by equation (2.29). The above transformation can be expressed in inhomogeneous coordinates as

$$ \begin{pmatrix} {}^{L,R}\hat{x} \\ {}^{L,R}\hat{y} \end{pmatrix} = \frac{f}{\widehat{Z}} \begin{pmatrix} \widehat{X} \pm b \\ \widehat{Y} \end{pmatrix} \tag{D.2} $$

taking the positive sign for L and the negative for R. Now, the optimal estimate is the point $\widehat{\mathbf{X}}$ that minimizes the image plane error between the measurements ${}^{L,R}\mathbf{x}$ and the projections ${}^{L,R}\hat{\mathbf{x}}$, calculated (in inhomogeneous coordinates) from equation (6.12) as:

$$ D^2 = |{}^{L}\hat{\mathbf{x}} - {}^{L}\mathbf{x}|^2 + |{}^{R}\hat{\mathbf{x}} - {}^{R}\mathbf{x}|^2 \tag{D.3} $$

Substituting equation (D.2) into equation (D.3):

$$ D^2 = \left(\frac{f(\widehat{X}+b)}{\widehat{Z}} - {}^{L}x \right)^2 + \left(\frac{f\widehat{Y}}{\widehat{Z}} - {}^{L}y \right)^2 + \left(\frac{f(\widehat{X}-b)}{\widehat{Z}} - {}^{R}x \right)^2 + \left(\frac{f\widehat{Y}}{\widehat{Z}} - {}^{R}y \right)^2 \tag{D.4} $$

Minimization of D^2 with respect to $\widehat{\mathbf{X}}$ proceeds in the usual manner by setting partial derivatives of D^2 to zero and solving for the optimal reconstruction. The relevant partial derivatives are:

$$\frac{\partial D^2}{\partial \widehat{X}} = \frac{2f}{\widehat{Z}}\left(\frac{f(\widehat{X}+b)}{\widehat{Z}} - {}^{L}x\right) + \frac{2f}{\widehat{Z}}\left(\frac{f(\widehat{X}-b)}{\widehat{Z}} - {}^{R}x\right) \tag{D.5}$$

$$\tag{D.6}$$

$$\frac{\partial D^2}{\partial \widehat{Y}} = \frac{2f}{\widehat{Z}}\left(\frac{f\widehat{Y}}{\widehat{Z}} - {}^{L}y\right) + \frac{2f}{\widehat{Z}}\left(\frac{f\widehat{Y}}{\widehat{Z}} - {}^{R}y\right) \tag{D.7}$$

$$\tag{D.8}$$

$$\begin{aligned}\frac{\partial D^2}{\partial \widehat{Z}} = &-\frac{2f(\widehat{X}+b)}{\widehat{Z}^2}\left(\frac{f(\widehat{X}+b)}{\widehat{Z}} - {}^{L}x\right) - \frac{2f\widehat{Y}}{\widehat{Z}^2}\left(\frac{f\widehat{Y}}{\widehat{Z}} - {}^{L}y\right) \\ &-\frac{2f(\widehat{X}-b)}{\widehat{Z}^2}\left(\frac{f(\widehat{X}-b)}{\widehat{Z}} - {}^{R}x\right) - \frac{2f\widehat{Y}}{\widehat{Z}^2}\left(\frac{f\widehat{Y}}{\widehat{Z}} - {}^{R}y\right)\end{aligned} \tag{D.9}$$

Setting the right hand side of equations (D.5) and (D.6) to zero and solving for $\widehat{X}$ and $\widehat{Y}$ gives:

$$\widehat{X} = \frac{\widehat{Z}}{2f}({}^{L}x + {}^{R}x) \tag{D.10}$$

$$\widehat{Y} = \frac{\widehat{Z}}{2f}({}^{L}y + {}^{R}y) \tag{D.11}$$

Now, setting $\frac{\partial D^2}{\partial \widehat{Z}} = 0$ in equation (D.8) and collecting common factors yields:

$$\begin{aligned}0 &= (\widehat{X}+b)(f(\widehat{X}+b) - {}^{L}x\widehat{Z}) + \widehat{Y}(f\widehat{Y} - {}^{L}y\widehat{Z}) \\ &\qquad + (\widehat{X}+b)(f(\widehat{X}-b) - {}^{R}x\widehat{Z}) + \widehat{Y}(f\widehat{Y} - {}^{R}y\widehat{Z}) \\ &= 2f\widehat{X}^2 + 2f\widehat{Y}^2 + 2fb^2 - \widehat{Z}\widehat{X}({}^{L}x + {}^{R}x) - \widehat{Z}\widehat{Y}({}^{L}y + {}^{R}y) - \widehat{Z}b({}^{L}x - {}^{R}x)\end{aligned} \tag{D.12}$$

Then, substituting equations (D.10)-(D.11) into equation (D.12):

$$\begin{aligned}0 &= \frac{\widehat{Z}^2}{2f}({}^{L}x + {}^{R}x)^2 + \frac{\widehat{Z}^2}{2f}({}^{L}y + {}^{R}y)^2 + 2fb^2 \\ &\qquad - \frac{\widehat{Z}^2}{2f}({}^{L}x + {}^{R}x)^2 - \frac{\widehat{Z}^2}{2f}({}^{L}y + {}^{R}y)^2 - \widehat{Z}b({}^{L}x - {}^{R}x) \\ &= 2fb - \widehat{Z}({}^{L}x - {}^{R}x)\end{aligned} \tag{D.13}$$

Finally, equation (D.13) can be solved for $\widehat{Z}$ and the result substituted into equations (D.10)-(D.11). Finally, the components of the optimal reconstruction $\widehat{\mathbf{X}}$ can be expressed as a function of the image plane measurements:

$$\widehat{X} = \frac{b}{{}^{L}x - {}^{R}x}({}^{L}x + {}^{R}x) \tag{D.14}$$

$$\widehat{Y} = \frac{b}{{}^{L}x - {}^{R}x}({}^{L}y + {}^{R}y) \tag{D.15}$$

$$\widehat{Z} = \frac{2bf}{{}^{L}x - {}^{R}x} \tag{D.16}$$

The above result can be compactly represented in vector form as:

$$\widehat{\mathbf{X}} = \frac{b}{{}^{L}x - {}^{R}x}\left({}^{L}x + {}^{R}x,\ {}^{L}y + {}^{R}y,\ 2f\right)^{\top} \tag{D.17}$$

which is the result quoted in equation (6.14) in Section 6.3.

D.2 Error Model for Reconstruction of a Single Point

This section now considers the error introduced in the reconstruction $\widehat{\mathbf{X}}$ (given by equation (D.17)) due to uncertainties in the parameters of the stereo camera model. Let b^*, f^* and ν^* represent the *actual* baseline, focal length and verge angle of the stereo rig, and let b, f and ν represent the *calibrated* parameters of the camera model. If the camera parameters are perfectly calibrated, the measurements ${}^{L,R}\mathbf{x}$ are related to the *actual* point, $\mathbf{X} = (X,Y,Z)^{\top}$, by (see equation(D.2)):

$${}^{L,R}\mathbf{x} = \frac{f^*}{Z}(X \pm 2b^*, Y)^{\top} \tag{D.18}$$

In practice, features are observed in non-rectilinear stereo cameras, and then corrected using projective rectification (see Section 2.2.4). However, any error in the calibrated verge angle will cause projective rectification to over or under-compensate by angle $(\nu - \nu^*)$ when correcting the raw measurements for rectilinear stereo. Thus, in the presence of verge angle error, the *ideal* rectilinear measurements in equation (D.18) are offset by angular error $(\nu - \nu^*)$ (see equations (2.37)-(2.38)) to give the *actual* rectilinear measurements:

$${}^{L,R}\mathbf{x} = \frac{f^*\left((X \pm 2b^*)\cos(\nu - \nu^*) \mp Z\sin(\nu - \nu^*),\ Y\right)^{\top}}{Z\cos(\nu - \nu^*) \pm (X \pm 2b^*)\sin(\nu - \nu^*)} \tag{D.19}$$

Assuming the verge angle error is not severe, the small angle approximations $\cos(\nu - \nu^*) \approx 1$ and $\sin(\nu - \nu^*) \approx (\nu - \nu^*)$ are introduced to reduce the above to

$${}^{L,R}\mathbf{x} = \frac{f^*\left((X \pm 2b^*) \mp Z(\nu - \nu^*),\ Y\right)^{\top}}{Z \pm (X \pm 2b^*)(\nu - \nu^*)} \tag{D.20}$$

Based on these measurements and *calibrated* camera parameters f, b and ν, the optimal reconstruction $\widehat{\mathbf{X}}$ is found by substituting equation (D.20) into equation (D.17). Evaluating the intermediate terms in equation (D.17):

$$
\begin{aligned}
{}^{L}x - {}^{R}x &= f^* \frac{(X+2b^*) - Z(\nu - \nu^*)}{Z + (X+2b^*)(\nu - \nu^*)} - f^* \frac{(X-2b^*) + Z(\nu - \nu^*)}{Z - (X-2b^*)(\nu - \nu^*)} \\
&= 2f^* \frac{2Zb^* - (X^2+Z^2-4(b^*)^2)(\nu - \nu^*) - 2Zb^*(\nu - \nu^*)^2}{Z^2 + 4Zb^*(\nu - \nu^*) - (X^2 - 4(b^*)^2)(\nu - \nu^*)^2}
\end{aligned} \tag{D.21}
$$

Again assuming the verge angle error is small, the approximation $(\nu - \nu^*)^2 \ll 1$ is introduced to reduce equation (D.21) to

$$
{}^{L}x - {}^{R}x = 2f^* \frac{2Zb^* - (X^2+Z^2-4(b^*)^2)(\nu - \nu^*)}{Z^2 + 4Zb^*(\nu - \nu^*)} \tag{D.22}
$$

The remaining intermediate terms follow similarly:

$$
{}^{L}x + {}^{R}x = 2f^* \frac{XZ}{Z^2 + 4Zb^*(\nu - \nu^*)} \tag{D.23}
$$

$$
{}^{L}y + {}^{R}y = 2f^* \frac{Y(Z + 2b^*(\nu - \nu^*))}{Z^2 + 4Zb^*(\nu - \nu^*)} \tag{D.24}
$$

Finally, the reconstruction $\widehat{\mathbf{X}} = (\widehat{X}, \widehat{Y}, \widehat{Z})^\top$ from equation (D.17) is

$$
\widehat{X} = \frac{XZb}{2Zb^* - (X^2+Z^2-4(b^*)^2)(\nu - \nu^*)} \tag{D.25}
$$

$$
\widehat{Y} = \frac{Yb(Z + 2b^*(\nu - \nu^*))}{2Zb^* - (X^2+Z^2-4(b^*)^2)(\nu - \nu^*)} \tag{D.26}
$$

$$
\widehat{Z} = \frac{2bf(Z^2 + 4Zb^*(\nu - \nu^*))}{2Zb^* - (X^2+Z^2-4(b^*)^2)(\nu - \nu^*)} \tag{D.27}
$$

Again assuming small errors, the effect of calibration errors on $\widehat{\mathbf{X}}$ is most easily examined by taking a Taylor series expansion of equations (D.25)-(D.27) with respect to f, b and ν about the operating point $f = f^*$, $b = b^*$ and $\nu = \nu^*$. Taking the first order expansion for each component:

$$
\widehat{X}_i(f,b,\nu) = X\left[1 + \frac{b-b^*}{2b^*} + \frac{X^2+Z^2-4(b^*)^2}{2Zb^*}(\nu - \nu^*)\right] \tag{D.28}
$$

$$
\widehat{Y}_i(f,b,\nu) = Y\left[1 + \frac{b-b^*}{2b^*} + \frac{X^2+Z^2}{2Zb^*}(\nu - \nu^*)\right] \tag{D.29}
$$

$$
\widehat{Z}_i(f,b,\nu) = Z\left[1 + \frac{b-b^*}{2b^*} + \frac{f-f^*}{f^*} + \frac{X^2+Z^2+4(b^*)^2}{2Zb^*}(\nu - \nu^*)\right] \tag{D.30}
$$

Rearrange the above and expressing in vector notation, the relationship between a real point $\mathbf{X} = (X,Y,Z)^\top$ and its reconstruction $\widehat{\mathbf{X}}$ in the presence of camera calibration errors can be written as a linear function of f, b and ν:

$$
\widehat{\mathbf{X}} = \left(1 + \frac{b-b^*}{2b^*} + \frac{X^2+Z^2}{2Zb^*}(\nu - \nu^*)\right)\begin{pmatrix} X \\ Y \\ Z \end{pmatrix} + \frac{f-f^*}{f^*}\begin{pmatrix} 0 \\ 0 \\ Z \end{pmatrix} + (\nu - \nu^*)\begin{pmatrix} 2Xb^*/Z \\ 0 \\ 2b^* \end{pmatrix} \tag{D.31}
$$

which is the final error model quoted in equation (6.17) in Section 6.3.

D.3 Error Model for Pose Estimation

The previous section considered the effect of calibration errors on the reconstruction of a single point. This section now examines the effect of errors on estimating the pose of an object modelled by multiple points. For simplicity and without loss of generality, the object is assumed to undergo pure translation. Let $\mathbf{G}_i$, $i = 1,\ldots,N$ represent the N points of the model, with $\sum \mathbf{G}_i = \mathbf{0}$ as discussed in Section 6.3, and let $\mathbf{T}_E = (X_E, Y_E, Z_E)^\top$ represent the *actual* position of the object. Now, let $\mathbf{G}_i^*$ represent the *visually reconstructed* points in the model. As discussed in Section 6.3, the effect of calibration errors is to scale each visually reconstructed point by K_1. Thus, the measured 3D points (after biasing by calibration errors) are effectively given by:

$$\mathbf{G}_i^* = K_1(\mathbf{G}_i + \mathbf{T}_E) \tag{D.32}$$

and the *actual* image plane measurements are given by the projection in equation (D.2):

$$^{L,R}\mathbf{g}_i = \frac{f}{K_1(Z_i + Z_E)}(K_1(X_i + X_E) \pm 2b,\ K_1(Y_i + Y_E))^\top \tag{D.33}$$

Now, let $\widehat{\mathbf{T}}_E = (\widehat{X}_E, \widehat{Y}_E, \widehat{Z}_E)^\top$ represent the *estimated* pose of the object. The points in the model are estimated as $\widehat{\mathbf{G}}_i = \mathbf{G}_i + \widehat{\mathbf{T}}_E$ (ie. without the unknown bias K_1) and the corresponding *predicted* measurements for the given pose are (from equation (D.2)):

$$^{L,R}\hat{\mathbf{g}}_i = \frac{f}{Z_i + \widehat{Z}_E}(X_i + \widehat{X}_E \pm 2b,\ Y_i + \widehat{Y}_E)^\top \tag{D.34}$$

An optimal estimate of the translation $\widehat{\mathbf{T}}_E$ is obtained by minimizing the reprojection error $D^2(\widehat{\mathbf{T}}_E)$ in equation (6.12) between the actual and predicted measurements. Substituting equations (D.33)-(D.34) into (6.12) gives the reprojection error as:

$$\begin{aligned}
D^2(\widehat{\mathbf{T}}_E) &= \sum_i \left[({}^L\hat{x}_i - {}^Lx_i)^2 + ({}^L\hat{y}_i - {}^Ly_i)^2 + ({}^R\hat{x}_i - {}^Rx_i)^2 + ({}^R\hat{y}_i - {}^Ry_i)^2\right] \\
&= f^2 \sum_i \left[\left(\frac{X_i + \widehat{X}_E + 2b}{Z_i + \widehat{Z}_E} - \frac{K_1(X_i + X_E) + 2b}{K_1(Z_i + Z_E)}\right)^2\right. \\
&\quad + \left(\frac{Y_i + \widehat{Y}_E}{Z_i + \widehat{Z}_E} - \frac{K_1(Y_i + Y_E)}{K_1(Z_i + Z_E)}\right)^2 \\
&\quad + \left(\frac{X_i + \widehat{X}_E - 2b}{Z_i + \widehat{Z}_E} - \frac{K_1(X_i + X_E) - 2b}{K_1(Z_i + Z_E)}\right)^2 \\
&\quad \left. + \left(\frac{Y_i + \widehat{Y}_E}{Z_i + \widehat{Z}_E} - \frac{K_1(Y_i + Y_E)}{K_1(Z_i + Z_E)}\right)^2\right]
\end{aligned}$$

The optimal estimated pose can be found analytically by setting the partial derivatives of $D^2(\widehat{\mathbf{T}}_E)$ to zero and solving for $\widehat{\mathbf{T}}_E$. Taking the partial derivative with respect to $\widehat{X}_E$ yields:

$$\frac{\partial D^2}{\partial \widehat{X}_E} = f^2 \sum_i \left[2 \left(\frac{X_i + \widehat{X}_E + 2b}{Z_i + \widehat{Z}_E} - \frac{K_1(X_i + X_E) + 2b}{K_1(Z_i + Z_E)} \right) \frac{1}{Z_i + \widehat{Z}_E} + 2 \left(\frac{X_i + \widehat{X}_E - 2b}{Z_i + \widehat{Z}_E} - \frac{K_1(X_i + X_E) - 2b}{K_1(Z_i + Z_E)} \right) \frac{1}{Z_i + \widehat{Z}_E} \right] \tag{D.35}$$

Expressing equation (D.35) over a common denominator and equating the numerator to zero gives

$$\begin{aligned} 0 = \sum_i \Big[& K_1(Z_i + Z_E)(X_i + \widehat{X}_E + 2b) - (Z_i + \widehat{Z}_E)[K_1(X_i + X_E) + 2b] \\ & + K_1(Z_i + Z_E)(X_i + \widehat{X}_E - 2b) - (Z_i + \widehat{Z}_E)[K_1(X_i + X_E) - 2b] \Big] \\ = & K_1 \sum (X_i Z_i) + K_1(\widehat{X}_E + 2b) \sum Z_i + K_1 Z_E \sum X_i + N K_1 Z_E (\widehat{X}_E + 2b) \\ & - K_1 \sum (X_i Z_i) - K_1(X_E + 2b) \sum Z_i - K_1 \widehat{Z}_E \sum X_i - N K_1 \widehat{Z}_E (X_E + 2b) \\ & + K_1 \sum (X_i Z_i) + K_1(\widehat{X}_E - 2b) \sum Z_i + K_1 Z_E \sum X_i + N K_1 Z_E (\widehat{X}_E - 2b) \\ & - K_1 \sum (X_i Z_i) - K_1(X_E - 2b) \sum Z_i - K_1 \widehat{Z}_E \sum X_i - N K_1 \widehat{Z}_E (X_E - 2b) \\ = & Z_E \widehat{X}_E - \widehat{Z}_E X_E \end{aligned}$$

where the last line is obtained after substituting $\sum \mathbf{G}_i = \mathbf{0}$. Solving for $\widehat{X}_E$:

$$\widehat{X}_E = \frac{X_E \widehat{Z}_E}{Z_E} \tag{D.36}$$

Taking the partial derivatives of $D^2(\widehat{\mathbf{T}}_E)$ with respect to $\widehat{Y}_E$, and following similar algebraic manipulation leads to

$$\widehat{Y}_E = \frac{Y_E \widehat{Z}_E}{Z_E} \tag{D.37}$$

and finally for $\widehat{Z}_E$:

$$\widehat{Z}_E = \frac{K_1[(\widehat{X}_E - X_E) \sum X_i Z_i + (\widehat{Y}_E - Y_E) \sum Y_i Z_i + Z_E (\sum X_i^2 + \sum Y_i^2 + N \widehat{X}_E^2 + N \widehat{Y}_E^2 + 4Nb^2)]}{K_1(\sum X_i^2 + \sum Y_i^2 + N X_E \widehat{X}_E + N Y_E \widehat{Y}_E) + 4Nb^2} \tag{D.38}$$

Substituting equations (D.36) and (D.37) into equation (D.38) allows $\widehat{X}_E$ and $\widehat{Y}_E$ to be eliminated from the expression for $\widehat{Z}_E$ (left as an exercise for the reader), and the result is substituted back into equations (D.36) and (D.37) to express the optimal estimated pose as a function of the real pose, the model points and the scale error K_1. The final solution can be expressed in vector notation as

$$\widehat{\mathbf{T}}_E = \frac{X_E \sum X_i Z_i + Y_E \sum Y_i Z_i - Z_E (\sum X_i^2 + \sum Y_i^2 + 4Nb^2)}{K_1(X_E \sum X_i Z_i + Y_E \sum Y_i Z_i) - Z_E[K_1(\sum X_i^2 + \sum Y_i^2) + 4Nb^2]} K_1 \mathbf{T}_E \tag{D.39}$$

which is the result quoted in equation (6.26) in Section 6.3.

E

Calibration of System Latencies

A service robot is a complex system requiring the coordination of multiple sensors and actuators. Complications arise when independent sensors sample measurements at different times, and actuators have an unknown delay between receiving and executing commands. Without compensation, these various latencies can have a negative impact on the stability and reliability of dynamic processes such as visual object tracking and feedback control. One way to avoid these effects is to ensure that the dynamics of the robot and environment remain within the region of stability. For the experiments in Chapters 6 and 7, for example, stability was ensured by setting a deliberately low gain on the visual servo controller. Obviously, the drawback of this approach is an increase in the total execution time. Other methods have been proposed to deal with latencies directly (for example, see [27]) for cases in which this side effect cannot be tolerated.

Clearly, methods that compensate for latency must possess good estimates of the relevant delays. This appendix presents a simple calibration procedure that can be used to estimate the latency between sensors and actuators. We illustrate this method by applying it to the light stripe sensor (see Chapter 3); specifically, estimating the delay between exposing the CCDs to a new image of the light stripe and sampling the angle encoder for the position of the light stripe. However, the same method also applies to the delay between sending a motion command to the Puma and observation of the subsequent motion on the CCD, or the delay between exposing the left and right CCDs (negligible on our experimental platform since the cameras are synchronized in hardware).

The calibration technique is based on minimizing the hysteresis induced by the delay for some cyclic motion. For the light stripe scanner, the laser is scanned backward and forward across a planar surface and the encoder value $e(t)$ and laser position on the image plane $x(t)$ (averaged over a few scan-lines) are recorded along with a timestamp t. Figure E.1 shows the image plane position plotted against encoder value, with a visible hysteresis loop resulting from the lag between these two variables. Now, the loop can be closed by phase shifting the encoder values by the acquisition delay δt. Thus, the latency δt may be calculated as the phase shift that minimizes the residual error from a linear regression applied to $x(t)$ and $e(t+\delta t)$. It

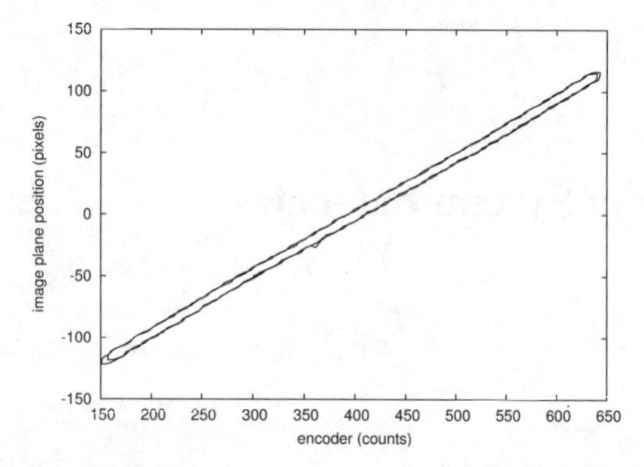

Fig. E.1. Hysteresis due to acquisition delay between corresponding images and encoder measurements of the light stripe.

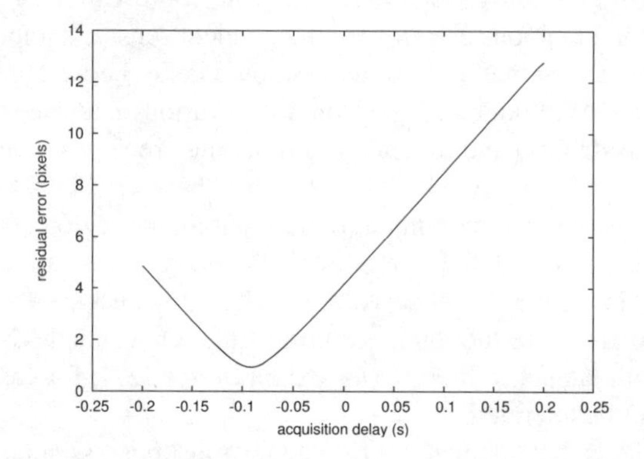

Fig. E.2. Residual error for estimation of time delay from linear regression.

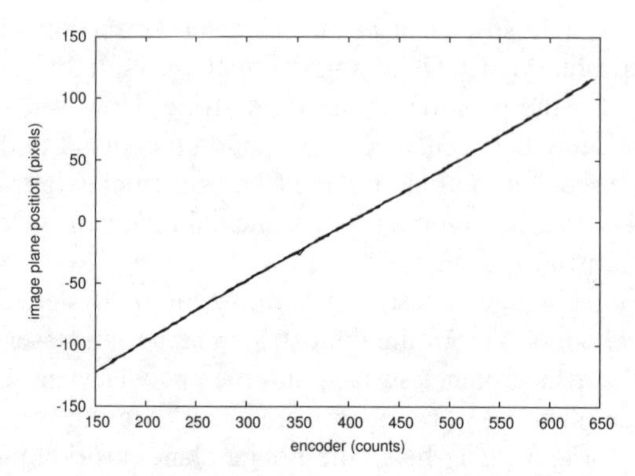

Fig. E.3. Relationship between captured images and encoder measurements of the light stripe after compensating for latency.

should be noted that the image plane position is not exactly a linear function of the encoder value, but the approximation is valid over a small range of motion.

The phase shift is applied using linear interpolation to calculate the encoder value $e(t+\delta t)$ corresponding to each image sample $x(t)$. Figure E.2 shows the residual error from a linear regression applied to $x(t)$ and $e(t+\delta t)$ for phase shifts in 5 ms increments between ± 200 ms. The plot exhibits a minima at about -90 ms (to the nearest 5 ms), indicating that the shaft encoder is sampled about 90 ms *after* the associated image was exposed. Finally, the relationship between encoder and CCD measurements after compensating for the delay is shown in Figure E.3, and the hysteresis has disappeared.

The calibration procedure was also applied to the Puma arm by moving the end-effector in a periodic motion and observing the image plane position of an active marker. The analysis revealed that the effect of a particular motion command was observed by the cameras about 170 ms after the command was issued to the Puma controller.

F

Task Planning

This appendix details the task planning required guide the motion of the end-effector for the experiments in Chapter 7. Task planning is divided into two phases: *grasp planning* and *trajectory planning*. Grasp planning is the process of determining the best contact points between the fingers and the target object to stably grasp the object. We simplify this process by formulating a specific algorithm for each known object, rather than using a general solution as in [87]. Based on established principles [37, 140], grasp planners for rectangular prisms (boxes) and upright cylinders (cups) are developed in Sections F.1 and F.2. Following grasp planning, a trajectory planner generates a collision-free path to guide the gripper from the current pose to the planned grasp. Details of the trajectory planner are presented in Section F.3. Finally, Section F.4 describes a simple inverse kinematic model for the Puma to calculate the angle of the wrist joint, which is required for grasp planning, and determine whether the a planned grasp is reachable.

F.1 Grasping a Box

While a box may be grasped in numerous ways, the algorithm described here simplifies the problem by considering only the precision grasp shown in the Figure F.1. This grasp is widely applicable and minimizes the possibility of collisions with other objects. Optimal grasp planning requires knowledge of surface properties such as friction the distribution of mass, both of which are typically unknown in service applications. However, general principles dictate that maximum stability is achieved when the force applied by the fingers is normal to the surface and the object is grasped near or above the centre of mass to minimize load torque when lifted. These rules are easily applied to a box, assuming a uniformly distributed mass.

Optimal grasp planning algorithms typically determine a set of candidate contact points on the surface of the object using heuristics and then select the best set based on a suitable cost function [38, 140]. The fast but sub-optimal planner described here considers only the two candidate contact pairs on orthogonal faces shown in Figure F.1: $(\mathbf{C}_1, \mathbf{C}_2)$ and $(\mathbf{C}_3, \mathbf{C}_4)$, where $\mathbf{C}_1$ and $\mathbf{C}_3$ are on the faces nearest the robot. The

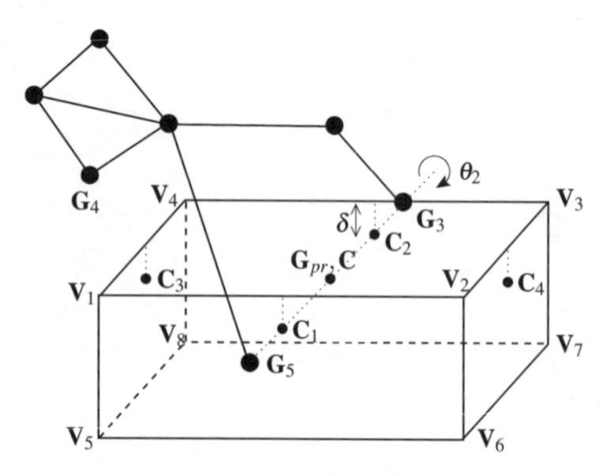

Fig. F.1. Grasp planning for a box.

contacts are placed a small distance δ below the top surface and above the midpoint of opposite faces to minimize torque and slippage. The contacts are calculated from the vertices $\mathbf{V}_1$ to $\mathbf{V}_8$ of the polygonal box model. For example, $\mathbf{C}_1$ is calculated as

$$\mathbf{C}_1 = \frac{1}{2}(\mathbf{V}_1 + \mathbf{V}_2) + \delta \cdot \frac{\mathbf{V}_5 - \mathbf{V}_1}{|\mathbf{V}_5 - \mathbf{V}_1|} \tag{F.1}$$

For clarity, the remaining discussion considers only $(\mathbf{C}_1, \mathbf{C}_2)$, and the equivalent calculations for $(\mathbf{C}_3, \mathbf{C}_4)$ follow similarly.

The first step in the planning process is to determine whether the contacts are reachable. Let $\mathbf{G}_5$ and $\mathbf{G}_3$ describe the location of the thumb and index fingertips when the gripper is fully opened. Candidates $(\mathbf{C}_1, \mathbf{C}_2)$ are considered to fit within the grasp when $|\mathbf{C}_1 - \mathbf{C}_2| < |\mathbf{G}_5 - \mathbf{G}_3|$. If the contact points violate this condition, the grasp candidate is considered no further. If successful, the algorithm calculates the transformation that aligns the fingertips with the contacts, by placing $\mathbf{G}_{pr}$ (defined in Figure 7.3) at the midpoint $\mathbf{C} = \frac{1}{2}(\mathbf{C}_1 + \mathbf{C}_2)$. The desired rotation aligns the pair of lines joining $(\mathbf{G}_5, \mathbf{G}_3)$ and $(\mathbf{C}_1, \mathbf{C}_2)$, and is calculated as an angle θ_1 about axis $\mathbf{A}_1$, given by:

$$\mathbf{A}_1 = \frac{({}^E\mathbf{G}_5 - {}^E\mathbf{G}_3) \times (\mathbf{C}_1 - \mathbf{C}_2)}{|{}^E\mathbf{G}_5 - {}^E\mathbf{G}_3||\mathbf{C}_1 - \mathbf{C}_2|} \tag{F.2}$$

$$\theta_1 = \cos^{-1}\left(\frac{({}^E\mathbf{G}_5 - {}^E\mathbf{G}_3)^\top (\mathbf{C}_1 - \mathbf{C}_2)}{|{}^E\mathbf{G}_5 - {}^E\mathbf{G}_3||\mathbf{C}_1 - \mathbf{C}_2|}\right) \tag{F.3}$$

The translation of the end-effector that aligns $\mathbf{G}_{pr}$ with $\mathbf{C}$ after rotation is

$$\mathbf{T}_1 = \mathbf{C} - \mathrm{R}_1(\theta_1, \mathbf{A}_1)^E\mathbf{G}_{pr} \tag{F.4}$$

where $\mathrm{R}_1(\theta_1, \mathbf{A}_1)$ is the rotation matrix corresponding to the axis/angle given in equations (F.2)-(F.3).

While the above transformation aligns the fingertips and contacts, the orientation of the hand about the line through $(\mathbf{C}_1, \mathbf{C}_2)$, represented as angle θ_2 in Figure F.1, remains unconstrained. To avoid collisions, the grasp planner chooses θ_2 such that the palm is above the top surface of the box. The lowest point on the gripper is $\mathbf{G}_4 = (X_4, Y_4, Z_4)^\top$, and the grasp is planned to satisfy $Y_4 > (Y_C + \delta + Y_{th})$, where Y_C is the height of $\mathbf{C}$ and error threshold Y_{th} ensures that $\mathbf{G}_4$ is well above the box. The location of $\mathbf{G}_4$ as a function of the orientation of the hand is:

$$^W\mathbf{G}_4(\theta_2) = \mathrm{R}_2(\theta_2, \mathbf{A}_2)\mathrm{R}_1(\theta_1, \mathbf{A}_1)(^E\mathbf{G}_4 - {}^E\mathbf{G}_{pr}) + \mathbf{C} \tag{F.5}$$

where the rotation axis is $\mathbf{A}_2 = (\mathbf{C}_2 - \mathbf{C}_1)/|\mathbf{C}_2 - \mathbf{C}_1|$. In the current implementation, θ_2 is calculated numerically as

$$\theta_2 = \mathrm{argmin}_\theta |Y_C + \delta + Y_{th} - Y_4(\theta)| \tag{F.6}$$

where $Y_4(\theta)$ is calculated from equation (F.5) for all angles in steps of one degree.

Finally, the two transformations are combined to give the planned orientation $^W\mathrm{R}_E$ and translation $^W\mathbf{T}_E$ of end-effector in the planned grasp:

$$^W\mathrm{R}_E = \mathrm{R}_2\mathrm{R}_1 \tag{F.7}$$

$$^W\mathbf{T}_E = \mathbf{C} - \mathrm{R}_2\mathrm{R}_1{}^E\mathbf{G}_{pr} \tag{F.8}$$

The condition developed in Section F.4 is used to determine whether the planned grasp is within reach of the robot, and the grasp is discarded if this condition is violated.

As noted earlier, a candidate grasp is calculated for both pairs of contact points, $(\mathbf{C}_1, \mathbf{C}_2)$ and $(\mathbf{C}_3, \mathbf{C}_4)$. If more than one grasp is realizable, the algorithm arbitrarily chooses the grasp that minimizes the wrist angle (Section F.4 describes the inverse kinematic model for calculating wrist angle of the Puma robot). Finally, the planned grasp is transformed to the frame of the end-effector as

$$^E\mathrm{H}_G = {}^W\mathrm{H}_E^{-1}{}^W\mathrm{H}_O \tag{F.9}$$

where $^W\mathrm{H}_O$ is the pose of the object, $^W\mathrm{H}_E$ is the planned pose of the end-effector given by equations (F.7)-(F.8), and $^E\mathrm{H}_G$ is defined in Figure 6.1. The rotational and translational components of $^E\mathrm{H}_G$ are:

$$^E\mathrm{R}_G = (\mathrm{R}_2\mathrm{R}_1)^{-1}{}^W\mathrm{R}_O \tag{F.10}$$

$$^E\mathbf{T}_G = {}^E\mathbf{G}_{pr} - (\mathrm{R}_2\mathrm{R}_1)^{-1}(^W\mathbf{T}_O + \mathbf{C}) \tag{F.11}$$

F.2 Grasping a Cup

Grasping for a cylindrical or conical cup follows a similar development to the previous section, except that a power grasp is employed since a precision grasp is difficult to stabilize on a curved surface. Only one candidate power grasp is possible for a cup, as shown in Figure F.2. The first step in planning is to determine whether the cup fits within the grasp, ie. $2r < |\mathbf{G}_5 - \mathbf{G}_3|$, where r is the radius of the cup and $\mathbf{G}_5$

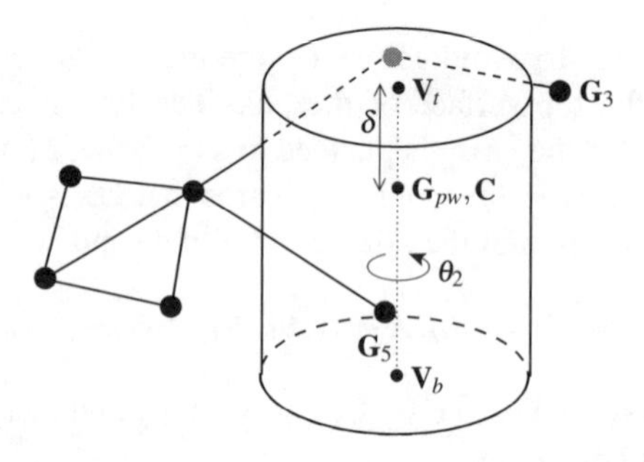

Fig. F.2. Grasp planning for an upright cup.

and $\mathbf{G}_3$ are the locations of the fingertips when the grasp is fully open. Furthermore, to ensure the hand does not collide with the floor, the height of the cup is required to satisfy $|\mathbf{V}_t - \mathbf{V}_b| > H_{th}$, where $\mathbf{V}_t$ and $\mathbf{V}_b$ (shown in Figure F.2) are the centroids of the top and bottom faces and H_{th} is the minimum height. For cups with sufficient height, the grasp centre $\mathbf{C}$ is positioned a small distance δ below the rim:

$$\mathbf{C} = \mathbf{V}_t + \delta \cdot \frac{\mathbf{V}_b - \mathbf{V}_t}{|\mathbf{V}_b - \mathbf{V}_t|} \tag{F.12}$$

To ensure the fingers exert a force approximately perpendicular to the surface, the hand is oriented to align the x-axis of the end-effector frame with the line joining $\mathbf{V}_t$ and $\mathbf{V}_b$. This alignment is achieved by rotating the end-effector by angle θ_1 about axis $\mathbf{A}_1$, given by:

$$\mathbf{A}_1 = \widehat{\mathbf{X}} \times (\mathbf{V}_t - \mathbf{V}_b)/|\mathbf{V}_t - \mathbf{V}_b| \tag{F.13}$$

$$\theta_1 = \cos(\widehat{\mathbf{X}}^\top (\mathbf{V}_t - \mathbf{V}_b)/|\mathbf{V}_t - \mathbf{V}_b|) \tag{F.14}$$

where $\widehat{\mathbf{X}} = (1,0,0)^\top$ is a unit vector in the direction of the x-axis.

To constrain the orientation of the hand about the axis of the cup (angle θ_2 in Figure F.2), the planning algorithm minimizes the angle of the wrist. Let θ_2 and $\mathbf{A}_2 = (\mathbf{V}_t - \mathbf{V}_b)/|\mathbf{V}_t - \mathbf{V}_b|$ represent the angle/axis of rotation about the axis of the cup. After rotation, the translation of the end-effector (aligning $\mathbf{C}$ and $\mathbf{G}_{pw}$) is:

$${}^W\mathbf{T}_E(\theta_2) = \mathbf{C} - \mathrm{R}_2(\theta_2, \mathbf{A}_2)\mathrm{R}_1(\theta_1, \mathbf{A}_1)^E\mathbf{G}_{pw} \tag{F.15}$$

Now, let $\theta_5({}^W\mathbf{T}_E)$ represent the angle of the wrist when the end-effector is positioned at ${}^W\mathbf{T}_E$ (details of wrist angle calculations are provided in Section F.4). The desired rotation that minimizes the wrist angle is given by:

$$\theta_2 = \mathrm{argmin}_\theta |\theta_5({}^W\mathbf{T}_E(\theta))| \tag{F.16}$$

As in the previous section, the minimization is performed as a numerical search over all angles in steps of one degree. Finally, the transformations are combined using equations (F.7)-(F.8) (with ${}^E\mathbf{G}_{pw}$ in place of ${}^E\mathbf{G}_{pr}$) to obtain the optimal pose of

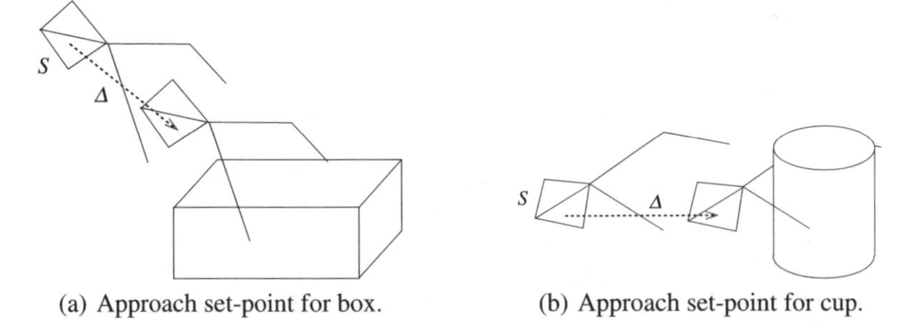

(a) Approach set-point for box. (b) Approach set-point for cup.

Fig. F.3. Trajectory planning for approaching the grasp of a box and cup.

the hand, ${}^{W}\mathrm{H}_{E}$. Finally, the reachability of the grasp is tested using the condition developed in Section (F.4). If this test is satisfied, the planned grasp is transformed to the end-effector frame ${}^{E}\mathrm{H}_{G}$ using equations (F.10)-(F.11). Otherwise, the cup is considered ungraspable.

F.3 Trajectory Planning

In general, a planned grasp cannot be approached from an arbitrary initial pose. Trajectory planning generates set-points to guide the motion of the end-effector along a suitable path to minimize the likelihood of collisions. A complete collision-free path planner should consider general obstacles, the target object itself and self-collisions. The planner developed in this section simplifies task planning by considering only collisions between the hand (or grasped object) and the target object. In this case, only a single set-point is required to plan a path from an arbitrary initial pose to the desired grasp (see Figure 7.4).

Figure F.3 shows the set-points generated for grasping a box and cup with the left hand. The target pose is approached away from the direction in which the fingers are pointing, which minimizes the chance of collision with the object. Let ${}^{W}\mathrm{R}_{S}$ and ${}^{W}\mathbf{T}_{S}$ represent the rotation and translation of the set-point S, and ${}^{W}\mathrm{R}_{E}$ and ${}^{W}\mathbf{T}_{E}$ represent the pose of the planned grasp. Then, the set-point (for a box or cup) is calculated as

$$ {}^{W}\mathbf{R}_{S} = {}^{W}\mathrm{R}_{E} \tag{F.17} $$
$$ {}^{W}\mathbf{T}_{S} = {}^{W}\mathbf{T}_{E} - \Delta\, {}^{W}\mathrm{R}_{E}\, {}^{E}\mathbf{F} \tag{F.18} $$

where ${}^{E}\mathbf{F}$ is the direction in which the fingers are pointing (in the end-effector frame E), and Δ is the distance between S and the final grasp. Using the planned set-point, the grasp is performed by visually servoing the hand first to S and then to the final pose.

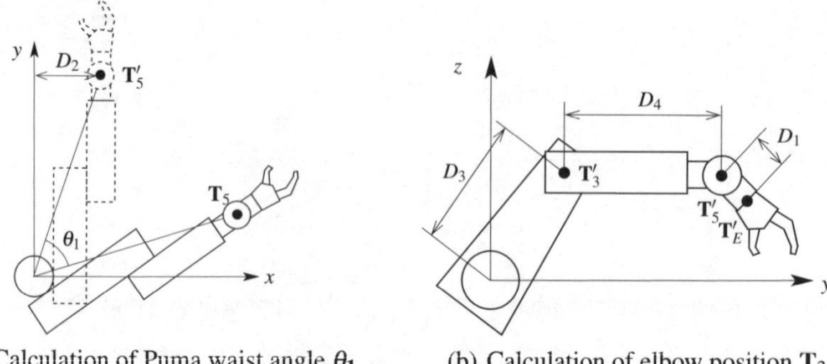

(a) Calculation of Puma waist angle θ_1 (b) Calculation of elbow position $\mathbf{T}_3$

Fig. F.4. Puma kinematic model.

F.4 Puma Kinematic Model

This section presents a simple inverse kinematics model for the Puma arm to recover the wrist angle (Puma joint 5) for a given pose of the end-effector, and also determine whether a planned pose is within the configuration space of the arm. These parameters are required by the grasp planning algorithms developed above.

Let $\mathbf{T}_E$ and R_E represent the position and orientation of the end-effector in the robot base frame, and let $\mathbf{T}_5$ represent the position of the wrist, which is distance D_1 from the origin of the end-effector frame (see Figure F.4). The location of the wrist can be calculated as

$$\mathbf{T}_5 = \mathbf{T}_E - D_1 \widehat{\mathbf{Z}}_E \tag{F.19}$$

where $\widehat{\mathbf{Z}}_E$ is a unit vector in the direction of the Z-axis of E, obtained from the third column of R_E. To calculate the position of the elbow (joint 3), the robot base frame is first rotated about its X-axis by angle θ_1 so that the arm is parallel to the YZ-plane, as shown in Figure F.4(a). The required angle of rotation is calculated from $\mathbf{T}_5 = (X_5, Y_5, Z_5)^\top$ as

$$\theta_1 = \tan^{-1}(Y_5/X_5) - \tan^{-1}\left(\sqrt{X_5^2 + Y_5^2 - D_2^2}/D_2\right) \tag{F.20}$$

After rotation, the position of the wrist is $\mathbf{T}_5' = (D_2, Y_5', Z_5')^\top$. Then, the position of the elbow $\mathbf{T}_3' = (D_2, Y_3', Z_3')^\top$ is at the intersection of two circles in the YZ-plane of radius D_3 and D_4, centred at the origin and $(Y_E', Z_E')^\top$ respectively (see Figure F.4(b)). The point of intersection has Y and Z-coordinates:

$$Z_3' = \frac{Z_5'(Y_5'^2 + Z_5'^2 + D_3^2 - D_4^2) + Y_5'\sqrt{4D_3^2(Y_5'^2 + Z_5'^2) - (Y_5'^2 + Z_5'^2 + D_3^2 - D_4^2)^2}}{2(Y_5'^2 + Z_5'^2)} \tag{F.21}$$

$$Y_3' = \sqrt{D_3^2 - Z_3'^2} \tag{F.22}$$

where the positive solutions are taken for the elbow pointing away from the body. Finally, the pose of the end-effector is beyond the reach of the Puma when the solution to equation (F.21) is imaginary, and the planned grasp in this case is unreachable. When the pose is reachable, the elbow position $\mathbf{T}'_3$ is rotated by $-\theta_1$ about the x-axis to give $\mathbf{T}_3$ in the original frame. Finally, the wrist angle $\theta_5(\mathbf{T}_E)$ for the desired pose is calculated as the angle between the lines joining $\mathbf{T}_5$ to $\mathbf{T}_3$ and $\mathbf{T}_5$ to $\mathbf{T}_E$:

$$\theta_5(\mathbf{T}_E) = \text{acos}\left(\frac{(\mathbf{T}_3 - \mathbf{T}_5)^\top(\mathbf{T}_E - \mathbf{T}_5)}{|\mathbf{T}_3 - \mathbf{T}_5||\mathbf{T}_E - \mathbf{T}_5|}\right) \tag{F.23}$$

The distances D_1 through to D_4 are taken from manufacturer specifications.

References

1. B. Adams, C. Breazeal, R. A. Brooks, and B. Scassellati. Humanoid robots: A new kind of tool. *IEEE Intelligent Systems*, 15(4):25–31, 2000.
2. J. I. Agbinya and D. Rees. Multi-object tracking in video. *Real-Time Imaging*, 5:295–304, 1999.
3. G. J. Agin and T. O. Binford. Computer description of curved objects. In *Proc. 3rd Int. Joint Conf. on Artifical Intelligence*, pages 629–640, 1973.
4. M. J. Aldon and L. Le Bris. Mobile robot localization using a light stripe sensor. In *Proc. of the Intelligent Vehicles '94 Symposium*, pages 255–259, 1994.
5. H. M. M. Alshawish and C. R. Allen. 3D object recognition using coded light projection for robot assembly applications. In *Proc. 21st International Conference on Industrial Electronics, Control and Instrumentation*, volume 2, pages 1420–1247, 1995.
6. N. Andreff, R. Horaud, and B. Espiau. Robot hand-eye calibration using structure from motion. *International Journal of Robotics Research*, 20(3):228–248, 2001.
7. G. C. Atkeson, J. G. Hale, F. Pollick, M. Riley, S. Kotosaka, S. Schaal, T. Shibata, G. Tevatia, A. Ude, S. Vijayakumar, E. Kawato, and M. Kawato. Using humanoid robots to study human behaviour. *IEEE Intelligent Systems*, 15(4):45–56, 2000.
8. Y. Bar-Shalom and X-R. Li. *Estimation and Tracking: Principles, Techniques and Software*. Artech House, 1993.
9. C. M. Bastuscheck. Techniques for real-time generation of range images. In *Proc. IEEE Comp. Soc. Conf. on Computer Vision and Pattern Recognition*, pages 262–268, 1989.
10. M. Becker, E. Kefalea, E. Maël, C. von der Malsburg, M. Pagel, J. Triesch, J. C. Vorbrüggen, R. P. Würtz, and S. Zadel. GripSee: A gesture-controlled robot for object perception and manipulation. *Autonomous Robots*, 6:203–21, 1999.
11. D.C. Bentivegna, A. Ude, C.G. Atkeson, and G. Cheng. Humanoid robot learning and game playing using PC-based vision. In *Proc. IEEE/RSJ 2002 International Conference on Intelligent Robots and Systems*, volume 3, pages 2449–2454, 2002.
12. P. J. Besl and R. C. Jain. Segmentation through variable-order surface fitting. *IEEE Transactions on Pattern Analysis and Machine Intelligence*, 10(2):167–192, March 1988.
13. I. Bloch. Information combination operators for data fusion: A comparative review with classification. *IEEE Transactions on Systems, Man and Cybernetics–Part A: Systems and Humans*, 26(1):52–67, 1996.
14. C. Breazeal. Socially intelligent robots: research, development, and applications. In *Proc. 2001 IEEE International Conference on Systems, Man, and Cybernetics*, volume 4, pages 2121–2126, 2001.

15. T. J. Broida and R. Chellappa. Estimation of object motion parameters from noisy images. *IEEE Transactions on Pattern Analysis and Machine Intelligence*, 8(1):90–99, Jan 1986.
16. R. A. Brooks. Intelligence without representation. *Artificial Intelligence Journal*, 47:139–159, 1991.
17. D. R. Butenhof. *Programming with POSIX threads*. Addison Wesley, 1997.
18. E. Cervera and P. Martinet. Visual servoing with indirect image control and a predictable camera trajectory. In *Proc. IEEE/RSJ International Conference on Intelligent Robots and Systems*, pages 381–386, 1999.
19. T. Chaperon and F. Goulette. Extracting cylinders in full 3D data using a random sampling method and the Gaussian image. In *Proc. Vision, Modeling and Visualization 2001*, pages 35–42, 2001.
20. F. Chaumette. Potential problems of stability and convergence in image-based and position-based visual servoing. In D. Kriegman, G. Hager, and A. Morse, editors, *The Confluence on Vision and Control*, volume 237 of *Lecture Notes in Control and Information Systems*, pages 66–78. Springer-Verlag, New York, 1988.
21. J. Chen and A. Zelinsky. Programming by demonstration: Coping with suboptimal teaching actions. *International Journal of Robotics Research*, 22(5):299–319, May 2003.
22. K. S. Chong. *Simultaneous Mapping and Localization for a Mobile Robot using Sonar Sensing*. PhD thesis, Monash University, Australia, 1997.
23. J-M. Chung and T. Nagata. Reasoning simplified volumetric shapes for robotic grasping. In *Proc. 1995 IEEE International Conference on Intelligent Robotics and Systems*, pages 348–353, 1995.
24. R. Cipolla and N. Hollinghurst. Visually guided grasping in unstructured environments. *Robotics and Autonomous Systems*, 19:337–346, 1997.
25. J. Clark, E. Trucco, and H-F. Cheung. Improving laser triangulation sensors using polarization. In *Proc. 5th Intnational Conference on Computer Vision*, pages 981–986, 1995.
26. F. S. Cohen and R. D. Rimey. A maximum likelihood approach to segmenting range data. In *Proc. 1988 IEEE International Conference on Robotics and Automation*, volume 3, pages 1696–1701, 1988.
27. P. I. Corke and M. C. Good. Dynamic effects in visual closed-loop systems. *IEEE Transactions on Robotics and Automation*, 12(5):671–683, 1996.
28. C. Dane and R. Bajcsy. Three-dimensional segmentation using the Gaussian image and spatial information. In *Proc. IEEE Computer Society Conference on Pattern Recognition and Image Processing*, pages 54–56, 1981.
29. T. Darrell, G. Gordon, M. Harville, and J. Woodfill. Integrated person tracking using stereo, color and pattern detection. *International Journal of Computer Vision*, 37(2):175–185, 2000.
30. A. Davison. *Mobile Robot Navigation Using Active Vision*. PhD thesis, University of Oxford, 1998.
31. L. Deng, W. J. Wilson, and F. Janabi-Sharifi. Characteristics of robot visual servoing methods and target model estimation. In *Proc. IEEE International Symposium on Intelligent Control*, pages 684–689, 2002.
32. M. Ehrenmann, R. Zollner, O. Rogalla, and R. Dillman. Programming service tasks in household environments by human demonstration. In *Proc. 11th IEEE International Workshop on Robot and Human Interactive Communication*, pages 460–467, 2002.

33. S. Ekvall and F. Hoffmann . Kragic. Object recognition and pose estimation for robotic manipulation using color cooccurrence histograms. In *Proc. IEEE/RSJ International Conference on Intelligent Robots and Systems*, pages 1284–1289, 2003.
34. B. Espiau, F. Chaumette, and P. Rives. A new approach to visual servoing in robotics. *IEEE Transactions on Robotics and Automation*, 8(3):313–326, 1992.
35. T-J. Fan, G. Medioni, and R. Nevatia. Segmented descriptions of 3-D surfaces. *IEEE Journal of Robotics and Automation*, 3(6):527–538, Dec 1987.
36. O. Faugeras. *Three-dimensional computer vision: a geometric viewpoint.* MIT Press, 1993.
37. C. Ferrari and J. Canny. Planning optimal grasps. In *Proc. 1992 IEEE International Conference on Robotics and Automation*, pages 2290–95, 1992.
38. M. Fischer and G. Hirzinger. Fast planning of precision grasps of 3D objects. In *Proc. IEEE/RSJ International Conference on Intelligent Robots and Systems*, pages 120–126, 1997.
39. D. A. Forsyth and J. Ponce. *Computer vision: a modern approach.* Prentice Hall, 2003.
40. H. Fujimoto, L.-C. Zhu, and K. Abdel-Malek. Image-based visual servoing for grasping unknown objects. In *Proc. 26th Annual Conference of the IEEE Industrial Electronics Society*, volume 2, pages 876–881, 2000.
41. V. Garric and M. Devy. Evaluation of calibration and localization methods for visually guided grasping. In *Proc. IEEE/RSJ International Conference on Intelligent Robots and Systems*, pages 387–393, 1995.
42. C. Gaskett, P. Brown, G. Cheng, and A. Zelinsky. Learning implicit models during target pursuit. In *Proc. IEEE International Conference on Robotics and Automation*, volume 3, pages 4122–4129, 2003.
43. Chris Gaskett and Gordon Cheng. Online learning of a motor map for humanoid robot reaching. In *Proceedings of the 2nd International Conference on Computational Intelligence, Robotics and Autonomous Systems*, 2003.
44. J. R. Goldschneider and A. Q. Li. Variational segmentation by peicewise facet models with application to range imagery. In *Proc. 2001 IEEE International Conference on Image Processing*, pages 812–815, 2001.
45. E. Grosso, G. Metta, A. Oddera, and G. Sandini. Robust visual servoing in 3-D reaching tasks. *IEEE Transactions on Robotics and Automation*, 12(5):732–742, 1996.
46. E. Grosso and G. Vercelli. Grasping strategies for reconstructed unknown 3D objects. In *Proc. IEEE/RSJ International Conference on Intelligent Robots and Systems*, pages 70–75, 1991.
47. A. Gruss, L. R. Carley, and T. Kanade. Integrated sensor and range-finding analog signal processor. *IEEE Journal of Solid-State Circuits*, 26(3):184–191, Mar 1991.
48. R. M. Haralick, J. Joo, C. Lee, X. Zhuang, V. G. Vaidya, and M. B. Kim. Pose estimation from corresponding point data. *IEEE Transactions on Systems, Man and Cybernetics*, 19(6):1426–1445, 1989.
49. R. M. Haralick, C. Lee, K. Ottenberg, and M. Nölle. Analysis and solutions of the three point perspective pose estimation problem. In *Proc. IEEE Conference on Computer Vision and Pattern Recognition*, pages 592–598, 1991.
50. R. M. Haralick, L. T. Watson, and T. J. Laffey. The topographic primal sketch. *The International Journal of Robotics Research*, 2(1):50–72, Spring 1983.
51. H. H. Harman. *Modern Factor Analysis*, chapter 8. The University of Chicago Press, 1969.
52. R. Hartley and A. Zisserman. *Multiple View Geometry in Computer Vision.* Cambridge University Press, 2000.

53. R. I. Hartley. Theory and practice of projective rectification. *International Journal of Computer Vision*, 35(2):115–127, 1999.
54. R. I. Hartley and P. Sturm. Triangulation. *Computer Vision and Image Understanding*, 68(2):146–157, 1997.
55. A. Hauck, J. Rüttinger, M. Sorg, and G. Färber. Visual determination of 3D grasping points on unknown objects with a binocular camera system. In *Proc. IEEE/RSJ International Conference on Intelligent Robots and Systems*, pages 272–278, 1999.
56. A. Hauck, M. Sorg, G. Faber, and T. Schenk. What can be learned from human reach-to-grasp movements from the design of robotic hand-eye systems. In *Proc. IEEE International Conference on Robotics and Automation*, pages 2521–2526, 1999.
57. J. Haverinen and J. Röning. An obstacle detection system using a light stripe identification based method. In *Proc. IEEE International Joint Symposium on Intelligence and Systems*, pages 232–236, 1998.
58. J. Haverinen and J. Röning. A 3-D scanner capturing range and colour for the robotics applications. In *24th Workshop of the AAPR*, May 25-26, Austria 2000.
59. M. Hebert. Active and passive range sensing for robotics. In *Proc. IEEE International Conference on Robotics and Automation*, pages 102–110, 2000.
60. D. D. Hoffman. *Visual Intelligence*, chapter 2. W. W. Norton and Company, 1998.
61. N. J. Hollinghurst. *Uncalibrated Stereo and Hand-Eye Coordination*. PhD thesis, University of Cambridge, Cambridge, 1997.
62. A. Hoover, G. Jean-Baptiste, X. Jiang, P. J. Flynn, H. Bunke, D. B. Goldgof, K. Bowyer, D. W. Eggert, A. Fitzgibbon, and R. B. Fisher. An experimental comparison of range image segmentation algorithms. *IEEE Transactions on Pattern Analysis and Machine Intelligence*, 18(7):673–689, July 1996.
63. R. Horaud and F. Dornaika. Visually guided object grasping. *IEEE Transactions on Robotics and Automation*, 14(4):525–532, 1998.
64. B. K. P Horn. Extended Gaussian images. *Proc. of the IEEE*, 72(12):1671–86, Dec 1984.
65. K. Hosoda and M. Asada. Versatile visual servoing without knowledge of the true Jacobian. In *Proc. IEEE/RSJ International Conference on Intelligent Robots and Systems*, pages 186–191, 1994.
66. C. Hu, M. Q. Meng, P. X. Liu, and X. Wang. Visual gesture recognition for human-machine interface of robot teleoperation. In *Proc. IEEE/RSJ International Conference on Intelligent Robots and Systems*, pages 1560–1565, 2003.
67. Y. Hu, R. Eagleson, and M. A. Goodale. Human visual servoing for reaching and grasping: The role of 3-D geometric features. In *Proc. IEEE International Conference on Robotics and Automation*, pages 3209–3216, 1999.
68. S. Hutchinson, G. D. Hager, and P. I. Corke. A tutorial on visual servo control. *IEEE Transactions on Robotics and Automation*, 12(5):651–670, 1996.
69. D. Q. Huynh, R. A. Owens, and P. E. Hartmann. Calibrating a structured light stripe system: A novel approach. *International Journal of Computer Vision*, 33(1):73–86, 1999.
70. K. Ikeuchi, M. Kawade, and T. Suehiro. Towards assembly plan from observation. In *Proc. IEEE/RSJ International Conference on Intelligent Robots and Systems*, pages 2294–2301, 1993.
71. *IA-32 Intel Architecture Software Developers Manual*. Intel Corporation, 2002.
72. M. Isard and A. Blake. ICONDENSATION: Unifying low-level and high-level tracking in a stochastic framework. In *Proc. 5th European Conference on Computer Vision*, volume 1, pages 893–908, 1998.
73. R. A. Jarvis. A perspective on range finding techniques for computer vision. *IEEE Transactions on Pattern Analysis and Machine Intelligence*, 5(2):122–139, 1983.

74. A. H. Jazwinski. *Stochastic Processes and Filtering Theory*. Mathematics in Science and Engineering. Academic Press, New York, 1970.
75. R. Joshi and A. C. Sanderson. Application of feature-based multi-view servoing for lamp filament alignment. *IEEE Robotics and Automation Magazine*, pages 25–31, December 1998.
76. R. E. Kahn, M. J. Swain, P. N. Prokopowicz, and R. J. Firby. Gesture recognition using the Perseus system. In *Proc. IEEE Computer Society Conference on Computer Vision and Pattern Recognition*, pages 734–741, 1996.
77. T. Kanade and D. D. Morris. Factorization methods for structure from motion. *Philosophical Transactions of the Royal Society of London, Series A*, 356(1740):1153–73, 1998.
78. T. Kanda, H. Ishiguro, T. Ono, M. Imai, and R. Nakatsu. Development and evaluation of an interactive humanoid robot "Robovie". In *Proc. 2002 IEEE International Conference on Robotics and Automation*, pages 1848–1855, 2002.
79. S. B. Kang and K. Ikeuchi. Determining 3-D object pose using the complex extended Gaussian image. In *Proc. IEEE Computer Society Conference on Computer Vision and Pattern Recognition*, pages 580–585, 1991.
80. E. Kefalea, E. Maël, and R. P. Würtz. An integrated object representation for recognition and grasping. In *Proc. 1999 Third International Conference on Knowledge-Based Intelligent Information Engineering Systems*, pages 423–426, 1999.
81. R. Kelly, R. Carelli, O. Nasisi, B. Kuchen, and F. Reyes. Stable visual servoing of camera-in-hand robotic systems. *IEEE Transactions on Mechatronics*, 5(1):39–48, 2000.
82. D. Khadraoui, G. Motyl, P. Martinet, J. Gallice, and F. Chaumette. Visual servoing in robotics scheme using a camera/laser-stripe sensor. *IEEE Transactions on Robotics and Automation*, 12(5):743–750, 1996.
83. H. Kim, J. Cho, and I. Kweon. A novel image-based control-law for the visual servoing system under large pose error. In *Proc. IEEE/RSJ International Conference on Intelligent Robots and Systems*, pages 263–267, 2000.
84. L. Kleeman. Optimal estimation of position and heading for mobile robots using ultrasonic beacons and dead-reckoning. In *IEEE International Conference on Robotics and Automation*, pages 2582–2587, 1992.
85. H. Kobayashi, K. Igawa, T. Bito, and K. Kikuchi. 3D object recognition and grasping for human support robotic systems. In *Proc. 2000 IEEE International Workshop on Robot and Human Interactive Communication*, pages 430–435, 2000.
86. A. Konno, T. Yoshiike, K. Nagashima, M. Inaba, and H. Inoue. Preliminary experiments in motion programming of humanoid robot by human demonstration. *JSME International Journal*, 43(2):401–407, 2000.
87. D. Kragić. *Visual Servoing for Manipulation: Robustness and Integration Issues*. PhD thesis, Royal Institute of Technology, Stockholm, 2001.
88. D. Kragic and H. I. Christensen. Model based techniques for robotic servoing and grasping. In *Proc. IEEE/RSJ International Conference on Intelligent Robots and Systems*, pages 299–304, 2002.
89. D. Kragić and H. I. Christensen. Survey on visual servoing for manipulation. Technical Report ISRN KTH/NA/P-02/01-SE, KTH, 2002.
90. P. Krsek, G. Lukács, and R. R. Martin. Algorithms for computing curvatures from range data. In *The Mathematics of Surfaces VIII*, pages 1–16, 1998.
91. A. Krupa, C. Doignon, J. Gangloff, and M. De Mathelin. Combining image-based and depth visual servoing applied to robotized laparoscopic surgery. In *IEEE/RSJ International Conference on Intelligent Robots and Systems*, pages 323–329, 2002.

92. J. J. Kuffner and S. M. LaValle. RRT-connect: An efficient approach to single-query path planning. In *Proc. IEEE International Conference on Robotics and Automation*, volume 2, pages 995–1001, 2000.
93. R. Langea, P. Seitza, A. Bibera, and R. Schwarteb. Time-of-flight range imaging with a custom solid-state image sensor. In *Proc. EOS/SPIEConference on Laser Metrology and Inspection*, volume SPIE 3823, pages 180–191, 1999.
94. J. M. Lavest, G. Rives, and M. Dhome. Three-dimensional reconstruction by zooming. *Proc. IEEE Transactions on Robotics and Automation*, 9(2):196–207, 1993.
95. C. S. Lovchik and M. A. Diftler. The Robonaut hand: A dextrous robot hand for space. In *Proc. IEEE International Conference on Robotics and Automation*, pages 907–912, 1999.
96. M. Magee, R. Weniger, and E. A. Franke. Location of features of known height in the presence of reflective and refractive noise using a stereoscopic light-striping approach. *Optical Engineering*, 33(4):1092–1098, April 1994.
97. E. Malis, F. Chaumette, and S. Boudet. 2-1/2-D visual servoing. *IEEE Transactions on Robotics and Automation*, 15(2):238–250, 1999.
98. E. Marchand, P. Bouthemy, F. Chaumette, and V. Moreau. Robust real-time visual tracking using a 2D-3D model-based approach. In *Proc. 7th IEEE International Conference on Computer Vision*, pages 262–268, 1999.
99. M. Marjanovic, B. Scassellati, and M. Williamson. Self-taught visually-guided pointing for a humanoid robot. In *Proc. Fourth International Conference on Simulation of Adaptive Behaviour*, pages 35–44, 1996.
100. D. Marshall, G. Lukacs, and R. Martin. Robust segmentation of primitives from range data in the presence of geometric degeneracy. *IEEE Transactions on Pattern Analysis and Machine Intelligence*, 23(3):304–314, March 2001.
101. S.J. McKenna, Y. Raja, and S. Gong. Object tracking using adaptive colour mixture models. In *Proc. Third Asian Conference on Computer Vision*, pages 615–622, 1997.
102. P. J. McKerrow. *Introduction to Robotics*. Addison-Wesley, 1991.
103. P. Michelman and P. Allen. Forming complex dextrous manipulations from task primitives. In *Proc. IEEE International Conference on Robotics and Automation*, volume 4, pages 3383–3388, 1994.
104. A. Morales, E. Chinellato, A. H. Fagg, and A. P. del Pobil. Experimental prediction of the performance of grasp tasks from visual features. In *Proc. IEEE/RSJ International Conference on Intelligent Robots and Systems*, pages 3423–3428, 2003.
105. J. J. Moré, B. S. Garbow, and K. E. Hillstrom. Users' guide for MINPACK-1. Technical Report ANL-80-74, Applied Math. Div., Argonne National Laboratory, 1980.
106. E. Mouaddib, J. Batllea, and J. Salvia. Recent progress in structured light in order to solve the correspondence problem in stereovision. In *Proc. IEEE International Conference on Robotics and Automation*, pages 130–136, 1997.
107. M. C. Moy. Gesture-based interaction with a pet robot. In *Proc. of the Sixteenth National Conference on Artificial Intelligence and Eleventh Conference on Innovative Applications of Artificial Intelligence*, pages 628–633, 1999.
108. M. Müller and H. Wörn. Planning of rapid grasp operations in unstructured scenes. In *Proc. 2000 IEEE/RSJ International Conference on Intelligent Robots and Systems*, volume 3, pages 1975–80, 2000.
109. Kouichi Nakano, Yasuo Watanabe, and Sukeyasu Kanno. Extraction and recognition of 3-dimensional information by projecting a pair of slit-ray beams. In *Proceedings of the 9th International Conference on Pattern Recognition*, pages 736–738, 1988.

110. K. Namba and N. Maru. 3D linear visual servoing for humanoid robot. In *28th Annual Conference of the IEEE Industrial Electronics Society*, volume 3, pages 2225–2230, 2002.
111. S. K. Nayar, A. C. Sanderson, L. E. Weiss, and D. A. Simon. Specular surface inspection using structured highlight and Gaussian image. *IEEE Transactions on Robotics and Automation*, 6(2):208–218, April 1990.
112. J. A. Nelder and R. Mead. A simplex method for function minimization. *Computer Journal*, 7:308–313, 1965.
113. B.J. Nelson and P.K. Khosla. An extendable framework for expectation-based visual servoing using environment models. In *Proc. IEEE International Conference on Robotics and Automation*, pages 184–189, 1995.
114. K. Nickels and S. Hutchinson. Model-based tracking of complex articulated objects. *IEEE Transactions on Robotics and Automation*, 17(1):28–36, 2001.
115. H. Nomura and T. Naito. Integrated visual servoing system to grasp industrial parts moving on conveyer by controlling 6DOF arm. In *Proc. IEEE/RSJ International Conference on Systems, Man and Cybernetics*, pages 1768–1775, 2000.
116. J. Nygards, T. Högström, and Å. Wernersson. Docking to pallets with feedback from a sheet-of-light range camera. In *Proc. 2000 IEEE/RSJ International Conference on Intelligent Robots and Systems*, pages 1853–1859, 2000.
117. J. Nygards and Å. Wernersson. Specular objects in range cameras: Reducing ambiguities by motion. In *Proc. of the IEEE International Conference on Multisensor Fusion and Integration for Intelligent Systems*, pages 320–328, 1994.
118. K. Okada, M. Inaba, and H. Inoue. Integration of real-time binocular stereo vision and whole body information for dynamic walking navigation of humanoid robot. In *Proc. IEEE Conference on Multisonsor Fusion and Integration for Intelligent Systems*, pages 131–136, 2003.
119. M. Okada, Y. Nakamura, and S. Ban. Design of programmable passive compliance shoulder mechanism. In *Proc. 2001 IEEE International Conference on Robotics and Automation*, volume 1, pages 348–353, 2001.
120. T. Okada, M. Sano, and H. Kaneko. Three-dimensional object recognition usind spherical correlation. In *Proc. 11th IAPR International Conference on Pattern Recognition*, volume 1, pages 250–254, 1992.
121. T. Okatani and K. Deguchi. Computation of the sign of the Gaussian curvature of a surface from multiple unknown illumination images without knowledge of the reflectance property. *Computer Vision and Image Understanding*, 76(2):125–134, November 1999.
122. V. I. Pavlovic, R. Sharma, and T. S. Huang. Visual interpretation of hand gestures for human-computer interaction: A review. *IEEE Transactions on Pattern Analysis and Machine Intelligence*, 19(7):677–695, 1997.
123. M. Pollefeys, L. Van Gool, M. Vergauwen, F. Verbiest, K. Cornelis, J. Tops, and R. Koch. Visual modeling with a hand-held camera. *International Journal of Computer Vision*, 59(3):207–232, 2004.
124. M. Pollefeys, R. Koch, and L. Van Gool. Self-calibration and metric reconstruction in spite of varying and unknown internal camera parameters. In *Proc. IEEE 6th International Conference on Computer Vision*, pages 90–95, 1998.
125. P. N. Prokopowicz, M. J. Swain, and R. E. Kahn. Task and environment-sensitive tracking. In *Proc. IARP/IEEE Workshop on Visual Behaviours*, pages 73–78, 1994.
126. K. Rao, G. Medioni, H. Liu, and G. A. Bekey. Shape description and grasping for robot hand-eye coordination. *IEEE Control Systems Magazine*, pages 22–29, February 1989.
127. R. P. N. Rao and D. H. Ballard. Learning saccadic eye movements using multiscale spatial filters. *Advances in Neural Information Processing Systems*, 7:893–900, 1995.

128. H. J. Ritter, T. M. Martinetz, and K. J. Schulten. Topology-conserving maps for learning visuo-motor-coordination. *Neural Networks*, 2(3):159–168, 1989.
129. P. Rives. Visual servoing based on epipolar geometry. In *Proc. IEEE/RSJ International Conference on Intelligent Robots and Systems*, pages 602–607, 2000.
130. M. A. Rodrigues, Y. F. Li, M. H. Lee, J. J. Rowland, and C. King. Robotic grasping of complex objects without full geometrical knowledge of shape. In *Proc. IEEE International Conference on Robotics and Automation*, pages 737–742, 1995.
131. O. Rogalla, M. Ehrenmann, R. Zollner, R. Becher, and R. Dillman. Using gesture and speech control for commanding a robot assistant. In *Proc. 11th IEEE International Workshop on Robot and Human Interactive Communication*, pages 454–459, 2002.
132. R. A. Russell and A. H. Purnamadjaja. Odor and airflow: complementary senses for a humanoid robot. In *Proc. IEEE International Conference on Robotics and Automation*, volume 2, pages 1842–1847, 2002.
133. R. A. Russell, G. Taylor, L. Kleeman, and A. H. Purnamadjaja. Multi-sensory synergies in humanoid robotics. *International Journal of Humanoid Robotics*, 1(2):289–314, 2004.
134. Y. K. Ryu and H. S. Cho. A neural network approach to extended Gaussian image based solder joint inspection. *Mechatronics*, 7(2):159–184, 1997.
135. A. C. Sanderson and L. E. Weiss. Image-based visual servo control using relational graph error signals. In *Proc. IEEE Conference on Cybernetics and Society*, pages 1074–1077, 1980.
136. B. Scassellati. A binocular, foveated active vision system. Technical Report 1628, MIT Artificial Intelligence Lab, 1998.
137. S. Schaal. Is imitation learning the route to humanoid robots. *Trends in Cognitive Sciences*, 3(6):233–242, 1999.
138. J. Shi and C. Tomasi. Good features to track. In *IEEE Conference on Computer Vision and Pattern Recognition*, pages 593–600, 1994.
139. Y. Shirai and M. Suwa. Recognition of polyhedrons with a range finder. In *Proc. 2nd Int. Joint Conf. on Artifical Intelligence*, pages 80–87, 1971.
140. G. Smith, E. Lee, K. Goldberg, K. Böhringer, and J. Craig. Computing parallel-jaw grips. In *Proc. 1999 IEEE International Conference on Robots and Automation*, pages 1897–1903, 1999.
141. M. Spengler and B. Schiele. Towards robust multi-cue integration for visual tracking. *Machine Vision and Applications*, 14:50–158, 2003.
142. J. Stavnitsky and D. Capson. Multiple camera model-based 3-D visual servo. *IEEE Transactions on Robotics and Automation*, 16(6):732–739, Dec 2000.
143. C. Sun and J. Sherrah. 3D symmetry detection using the extended Gaussian image. *IEEE Transactions on Pattern Analysis and Machine Intelligence*, 19(2):164–169, February 1997.
144. A. Takanishi, M. Ishida, Y. Yamazaki, and I. Kato. The realization of dynamic walking by the biped walking robot wl-10rd. In *Proc. '85 International Conference on Advanced Robotics*, pages 459–466, 1985.
145. A. Takanishi, H. Lim, M. Tsuda, and I. Kato. Realization of dynamic biped walking stabilized by trunk motion on a sagittally uneven surface. In *Proc. IEEE International Workshop on Intelligent Robots and Systems*, pages 323–330, 1990.
146. M. Takizawa, Y. Makihara, N. Shimada, J. Miura, and Y. Shirai. A service robot with interactive vision – object recognition using dialog with user –. In *Fist International Workshop on Language Understanding and Agents for Real World Interaction*, pages 16–23, 2003.

147. G. Taylor and L. Kleeman. Grasping unknown objects with a humanoid robot. In *Proc. 2002 Australiasian Conference on Robotics and Automation*, pages 191–196, 2002.
148. G. Taylor and L. Kleeman. Robust range data segmentation using geometric primitives for robotic applications. In *Proc. 9th IASTED International Conference on Signal and Image Processing*, pages 467–472, 2003.
149. G. Taylor and L. Kleeman. Hybrid position-based visual servoing with online calibration for a humanoid robot. In *Proc. IEEE/RSJ International Conference on Intelligent Robots and Systems*, pages 686–691, 2004.
150. G. Taylor and L. Kleeman. Integration of robust visual perception and control for a domestic humanoid robot. In *Proc. IEEE/RSJ International Conference on Intelligent Robots and Systems*, pages 1010–1015, 2004.
151. G. Taylor and L. Kleeman. Stereoscopic light stripe scanning: Interference rejection, error minimization and calibration. *International Journal of Robotics Research*, 23(12):1141–1156, December 2004.
152. M. Tonko, K. Schäfer, F. Heimes, and H.-H. Nagel. Towards visually servoed manipulation of car engine parts. In *Proc. IEEE International Conference on Robotics and Automation*, pages 1366–1371, 1997.
153. K. Toyama and G.D. Hager. Incremental focus of attention for robust vision-based tracking. *International Journal of Computer Vision*, 35(1):45–63, 1999.
154. E. Trucco and R. B. Fisher. Computing surface-based representations rrom range images. In *Proc. IEEE International Symposium on Intelligent Control*, pages 275–280, 1992.
155. E. Trucco and R. B. Fisher. Acquisition of consistent range data using local calibration. In *Proc. of the 1994 IEEE International Conference on Robotics and Automation*, volume 4, pages 3410–3415, 1994.
156. R. Y. Tsai and R. K. Lenz. A new technique for fully autonomous and efficient 3D robotic hand/eye calibration. *IEEE Transactions on Robotics and Automation*, 5(3):345–358, 1989.
157. T. Uhlin, P. Nordlund, A. Maki, and J.-O. Eklundh. Towards an active visual observer. In *Proc. of the International Conference on Computer Vision*, pages 679–686, Cambridge, MA, 1995.
158. K. Usher, P. Ridley, and P. Corke. Visual servoing for a car-like vehicle - an application in omnidirectional vision. In *Proc. 2002 Australasian Conference on Robotics and Automation*, pages 37–42, 2002.
159. S. van der Zwann, A. Bernardino, and J. Santos-Victor. Vision based station keeping and docking for an aerial blimp. In *IEEE/RSJ International Conference on Intelligent Robots and Systems*, volume 1, pages 614–619, 2000.
160. P. Viola and M. J. Jones. Robust real-time face detection. *International Journal of Computer Vision*, 57(2):137–154, 2004.
161. G. Welch and G. Bishop. SCAAT: Incremental tracking with incomplete information. *Computer Graphics*, 31:333–344, 1997.
162. D. S. Wheeler, A. H. Fagg, and R. A. Grupen. Learning prospective pick and place behaviour. In *Proc. The 2nd International Conference on Development and Learning*, pages 197–202, 2002.
163. W. Wilson, C. Williams Hulls, and G. Bell. Relative end-effector control using cartesian position based visual servoing. *IEEE Transactions on Robotics and Automation*, 12(5):684–696, 1996.
164. P. Wira and J. P. Urban. A new adaptive Kalman filter applied to visual servoing tasks. In *Fourth International Conference on Knowledge-Based Intelligent Engineering Systems and Applied Technologies*, pages 267–270, 2000.

165. P. Wunsch and G. Hirzinger. Real-time visual tracking of 3-D objects with dynamic handling of occlusion. In *Proc. IEEE International Conference on Robotics and Automation*, pages 2868–2873, 1997.
166. J. Vanden Wyngaerd and L. Van Gool. Coarse registration of surface patches with local symmetries. In *Proc. 7th European Conference on Computer Vision*, pages 572–586, 2002.
167. H. S. Yang and A. C. Kak. Determination of the identity, position and orientation of the topmost object in a pile: Some futher experiments. In *Proc. 1986 IEEE International Conference on Robotics and Automation*, volume 1, pages 293–298, 1986.
168. Y. Yokokohji, M. Sakamoto, and T. Yoshikawa. Vision-aided object manipulation by a multifingered hand with soft fingertips. In *Proc. IEEE International Conference on Robotics and Automation*, pages 3201–3208, 1999.

Index